The series "Studies in Computational Intelligence" (SCI) publishes new developments and advances in the various areas of computational intelligence—quickly and with a high quality. The intent is to cover the theory, applications, and design methods of computational intelligence, as embedded in the fields of engineering, computer science, physics and life sciences, as well as the methodologies behind them. The series contains monographs, lecture notes and edited volumes in computational intelligence spanning the areas of neural networks, connectionist systems, genetic algorithms, evolutionary computation, artificial intelligence, cellular automata, self-organizing systems, soft computing, fuzzy systems, and hybrid intelligent systems. Of particular value to both the contributors and the readership are the short publication timeframe and the world-wide distribution, which enable both wide and rapid dissemination of research output.

More information about this series at http://www.springer.com/series/7092

Jagdish Chand Bansal · Pramod Kumar Singh
Nikhil R. Pal
Editors

Evolutionary and Swarm Intelligence Algorithms

 Springer

Editors
Jagdish Chand Bansal
South Asian University
New Delhi, Delhi
India

Pramod Kumar Singh
ABV-Indian Institute of Information
   Technology and Management
Gwalior, Madhya Pradesh
India

Nikhil R. Pal
Electronics and Communication Sciences
   Unit (ECSU)
Indian Statistical Institute
Kolkata, West Bengal
India

ISSN 1860-949X   ISSN 1860-9503   (electronic)
Studies in Computational Intelligence
ISBN 978-3-030-08229-1   ISBN 978-3-319-91341-4   (eBook)
https://doi.org/10.1007/978-3-319-91341-4

Preface

Due to the high complexity of real-world optimization problems, often it is not easy to solve them using traditional or deterministic optimization methods. There are many real-world optimization problems for which one can afford near-optimal solution rather than an exact solution. Therefore, a class of robust algorithms is required, which does not depend upon the particular characteristics of the problems and hence can be applied to a wide variety of problems. Evolutionary computation and swarm intelligence-based optimization algorithms serve the purpose. Swarm and evolutionary algorithms are probabilistic algorithms, which are often very effective with problems that are not easy to deal with classical optimization methods. However, we want to emphasize that it is not our intention to say that these families provide a set of all-cure solutions. In fact, because of the stochastic nature of the search process, reproducibility may become a challenging issue unless one is careful about the experiments. Often, the computational overhead could be very high also. If a problem can be tackled with a classical optimization method for which the characteristics of the solutions can be analyzed easily, our prescription is not to use swarm or evolutionary algorithms for such a problem.

This book provides a detailed study and working procedure of few algorithms in the area of swarm intelligence and evolutionary computation. It contains totally nine chapters on various swarm and evolutionary algorithms and their recent application areas. Chapters "Swarm and Evolutionary Computation" to "Spider Monkey Optimization Algorithm" deal with swarm intelligence; Chapters "Genetic Algorithm and Its Advances in Embracing Memetics" and "Constrained Multi-objective Evolutionary Algorithm" focus on genetic algorithms and evolutionary multi-objective optimization, while Chapters "Genetic Programming for Classification and Feature Selection" to "Evolutionary Fuzzy Systems: A Case Study for Intrusion Detection Systems" are on genetic programming.

Chapter "Swarm and Evolutionary Computation" provides a detailed introduction to the two families of algorithms: swarm intelligence and evolutionary computation. This chapter also makes a comparative discussion of the two families and presents their advantages and limitations. Chapter "Particle Swarm Optimization" presents one of the most significant swarm intelligence-based algorithms, Particle Swarm

Optimization (PSO). Apart from its working mechanism, this chapter also explains the significance of each term of the update equation in PSO. Artificial Bee Colony (ABC) optimization algorithm, another very popular member of the swarm intelligence family, is discussed in Chapter "Artificial Bee Colony Algorithm Variants and Its Application to Colormap Quantization". This chapter details the ABC for constrained, multi-objective, and combinatorial optimization problems. ABC has also been applied to colormap quantization problem. In Chapter "Spider Monkey Optimization Algorithm", the Spider Monkey Optimization (SMO), a relatively new member of the swarm intelligence family is introduced. The SMO is a fission-fusion social structure-based optimization algorithm. The chapter explains the motivation and the detailed working mechanism along with a numerical example.

Chapter "Genetic Algorithm and Its Advances in Embracing Memetics" deals with genetic algorithms, particularly genetic algorithms with memetics. Authors first consider meme as a local search process, or an individual learning procedure, whose intensity can be governed by a theoretically derived upper bound. Then, they also treat meme as a building block of structured knowledge, which can be learned and transferred across problem instances for more efficient search. Genetic algorithms with memetics are applied to solve NP-hard capacitated arc routing problem. Evolutionary bilevel optimization is also discussed briefly in this chapter. Evolutionary multi-objective optimization (EMO) algorithms that are specifically designed for handling constraints are discussed in Chapter "Constrained Multiobjective Evolutionary Algorithm". Some numerical test problems as well some engineering design problems involving constraints are discussed. The authors provide a number of future research directions in the field of EMO also. Chapter "Genetic Programming for Classification and Feature Selection" presents a detailed application of the evolutionary fuzzy systems on intrusion detection. Evolutionary fuzzy systems is a generalization of genetic fuzzy systems. Apart from the taxonomy of evolutionary fuzzy systems, the chapter very nicely explains every step of generating an evolutionary fuzzy system. Finally, an application to intrusion detection is presented.

The remaining two chapters cover different aspects of genetic programming. Chapter "Genetic Programming for Job Shop Scheduling" focuses on genetic programming (GP). Naive model of GP-based binary classification strategy is provided. The chapter also discusses important issues related to GP when it is used for classification and feature selection. In Chapter "Evolutionary Fuzzy Systems: A Case Study for Intrusion Detection Systems," authors present an interesting application of genetic programming, job shop scheduling (JSS), a difficult operations research problem. This chapter also provides a brief survey of studies on the dispatching rules for job shop scheduling. Ideas to improve GP for job shop scheduling are also presented.

New Delhi, India   Jagdish Chand Bansal
Gwalior, India   Pramod Kumar Singh
Kolkata, India   Nikhil R. Pal

Contents

About the Editors

Jagdish Chand Bansal is Assistant Professor in the Department of Mathematics of the South Asian University, Delhi, India. His current research includes swarm intelligence, evolutionary algorithms, and optimization. He is Editor-in-Chief of the *International Journal of Swarm Intelligence*. He has given many keynote and invited talks in different international conferences in the area of soft computing. He is General Secretary of Soft Computing Research Society.

Pramod Kumar Singh completed his B.Tech. (CSE) from KNIT Sultanpur, and M.Tech. (CSE) and Ph.D. (CSE) from IIT Kharagpur. Currently, he is associated with ABV-Indian Institute of Information Technology and Management, Gwalior (ABV-IIITM, Gwalior) as Associate Professor. Prior to this, he has been associated with NIT Jalandhar as Lecturer and Senior Lecturer, SLIET Longowal as Assistant Professor, and IIT Kharagpur as Networking Engineer and Senior Networking Engineer. His research interests are nature-inspired computing, multi-objective optimization, data mining, text mining, and machine learning.

Nikhil R. Pal is Professor in the Electronics and Communication Sciences Unit of the Indian Statistical Institute. His current research interest includes brain science, computational intelligence, machine learning, and data mining.

He was Editor-in-Chief of the *IEEE Transactions on Fuzzy Systems* for the period January 2005–December 2010. He has served/been serving on the editorial/advisory board/steering committee of several journals including the *International Journal of Approximate Reasoning*, *Applied Soft Computing*, *International Journal of Neural Systems*, *Fuzzy Sets and Systems*, *IEEE Transactions on Fuzzy Systems*, and the *IEEE Transactions on Cybernetics*.

He is a recipient of the 2015 IEEE Computational Intelligence Society (CIS) Fuzzy Systems Pioneer Award, and he has given many plenary/keynote speeches in different premier international conferences in the area of computational intelligence. He is Distinguished Lecturer of the IEEE CIS (2010–2012, 2016–2018) and was Member of the Administrative Committee of the IEEE CIS (2010–2012). He has served as Vice President for Publications of the IEEE CIS (2013–2016). He is serving as President of the IEEE CIS (2018–2019).

He is Fellow of the National Academy of Sciences, India, Indian National Academy of Engineering, Indian National Science Academy, International Fuzzy Systems Association (IFSA), The World Academy of Sciences, and Fellow of the IEEE, USA.

Swarm and Evolutionary Computation

Jagdish Chand Bansal and Nikhil R. Pal

Abstract Optimization problems arise in various fields of science, engineering, and industry. In many occasions, such optimization problems, particularly in the present scenario, involve a variety of decision variables and complex structured objectives, and constraints. Often, the classical or traditional optimization techniques face difficulty in solving such real world optimization problems in their original form. Due to deficiencies of classical optimization algorithms in solving large-scale, highly non-linear, and often non-differentiable problems, there is a need to develop efficient and robust computational algorithms, which can solve problems, numerically irrespective of their sizes. Taking inspiration from nature to develop computationally efficient algorithms is one way to deal with real world optimization problems. Broadly, one can put these algorithms in the field of computational sciences and in particular, to computational intelligence. Formally, *computational intelligence* (CI) is a set of nature-inspired computational methodologies and approaches to solve complex real world problems. The major constituents of CI are Fuzzy Systems (FS), Neural Networks (NN), and *Swarm Intelligence (SI)* and *Evolutionary Computation (EC)*. Computational intelligence techniques are powerful, efficient, flexible, and reliable. *Swarm Intelligence* and *Evolutionary Computation* are two very useful components of computational intelligence that are primarily used to solve optimization problems. This book primarily concerns with various swarm and evolutionary optimization algorithms. This chapter provides a brief introduction to swarm and evolutionary algorithms.

Keywords Swarm intelligence · Evolutionary computation · Computational intelligence · Optimization

J. C. Bansal (✉)
Department of Mathematics, South Asian University, New Delhi, India
e-mail: jcbansal@gmail.com

N. R. Pal
Electronics and Communication Sciences Unit (ECSU),
Indian Statistical Institute, Kolkata, India
e-mail: nikhil@isical.ac.in

© Springer International Publishing AG, part of Springer Nature 2019   1
J. C. Bansal et al. (eds.), *Evolutionary and Swarm Intelligence*
Algorithms, Studies in Computational Intelligence 779,
https://doi.org/10.1007/978-3-319-91341-4_1

1 Swarm Intelligence

The word *swarm* refers to a collection of disorganized moving individuals or objects like insects, birds, fishes. More formally, a *swarm* can be considered a collection of interacting homogeneous agents or individuals. Researchers have developed many useful algorithms by modeling and simulating the foraging behavior of these individuals. The term Swarm Intelligence was coined by Beni and Wang in connection with cellular robotic systems [3]. They developed a set of algorithms for controlling robotic swarms. However, there is at least one earlier work (there may be more) that exploited flocking behavior of birds. For example, in 1987 Reynolds [17] developed a program to simulate motion of flocks of birds or animals using individual behavior.

Swarm intelligence is a discipline that deals with natural and artificial systems composed of many individuals that coordinate based on the decentralized, collective and self-organized cooperative behavior of social entities like flock of birds, or school of fishes, ant colonies, animal herding, bacterial growth, and microbial intelligence. The members of a swarm must be active, dynamic and simple (with no or very little inherent knowledge of the surroundings). Within the swarm, due to this cooperative behavior, a search strategy, better than random search, emerges. The so obtained intelligent search strategy may be referred to as swarm intelligence, in general. A well-accepted definition of swarm intelligence is by Bonabeau et al. [5], The emergent collective intelligence of groups of simple agents.

Following the early work in the late 80s, in the 90s two successful algorithms, named Ant colony optimization (in 1992) [8] and Particle swarm optimization (in 1995) [15] were developed. Till the mid-90s swarm intelligence approach was considered under evolutionary computation approaches due to their similarities in terms of the use of population, stochastic nature, and fields of application. However, nowadays, SI got its own identity because of some inherent differences between the underlying philosophies of SI and EC. SI attempts to mimic collective and synergistic behavior of simple agents, while EC is inspired by biological evolution. Because of its simplicity and effectiveness in solving real world problems, SI has become quite popular as a class of optimization algorithms.

The algorithms in the class of swarm intelligence, primarily consists of two phases, namely the *variation phase* and the *selection phase*. These phases are responsible for maintaining the balance between exploration and exploitation and to force the entire swarm; i.e., the set of potential solutions, to update their positions. The variation phase explores different areas of the search space and the selection phase works for the exploitation of the previous experiences.

Karaboga [14] presented the necessary and sufficient conditions for swarm intelligence. According to Karaboga, a group of homogeneous agents exhibits the swarm intelligence if and only if it follows two conditions: self-organization and division of labor.

1.1 Self-Organization

It is a process where some ordered movement arises from the local interaction between individuals of an initially disordered group. Bonabeau et al. [5] categorized self-organization in four strategies:

- **Positive feedback**: This is an information extracted from the output system and is revealed to the input system to promote formation of appropriate structures. Positive feedback provides diversity in swarm intelligence.
- **Negative feedback**: It balances the positive feedback and provides stabilization to the collective pattern. Negative feedback refers to exploitation in swarm intelligence.
- **Fluctuations**: These are related to the randomness in the system. Fluctuations provide new situations in the process and help to get rid of stagnation.
- **Multiple interaction**: This provides the way of learning from more than one individual within a society and improves the overall intelligence of the swarm.

1.2 Division of Labor

This helps different tasks to be performed simultaneously by different specialized individuals. Division of labor makes the swarm capable of handling changed conditions in the search space. A list of popular and successful swarm intelligence algorithms is:

- Particle Swarm Optimization [15]
- Spider Monkey Optimization [1]
- Artificial Bee Colony Algorithm [14]
- Ant Colony Optimization [8]
- Bacterial Foraging Optimization [16]
- Firefly algorithm [18].

The Particle Swarm Optimization (PSO) is inspired by birds flocking or fish schooling, the Artificial Bee Colony (ABC) is motivated by foraging behavior of honey bees, while the Ant Colony Optimization (ACO) is inspired by foraging behavior of ants. The Bacterial Foraging Optimization (BFO), on the other hand, is inspired by the group foraging behavior of bacteria such as E.coli and M. xanthus; the Firefly Algorithm (FFA) drew its inspiration from the flashing behavior of fireflies, while the Spider Monkey Optimization (SMO) is guided by foraging behavior of spider monkeys.

Giving a detailed overview of all these algorithms is beyond the scope of this Chapter. A quick search through literature reveals that of these algorithms, PSO and ACO are the most popular ones. Hence, we very briefly discuss these two algorithms.

Particle Swarm Optimization (PSO) is inspired by the social behavior of flock of birds or school of fishes. It is developed by Kennedy and Eberhart [15] in 1995.

Typically, a flock of birds have no leader and they find food by collaborative trial and error. They follow one of the members of the group that has the closest position with the food source. Others update their position simultaneously by communicating with members who already have a better position. This is done repeatedly until the best food source is found. Particle swarm optimization consists of a population of individuals called swarm, and each individual is called a particle, which represents a location or possible candidate solution in a multidimensional search space. Refer Chap. 6 for more details about PSO.

Ant Colony Optimization or ACO [8] is the first successful example of swarm intelligence. The algorithm was introduced to find the optimal path in a graph. Ant Colony Optimization algorithm is inspired from ants ability to find the shortest path between their nest and food source (Refer Fig. 1). A group of ants starts food foraging randomly. Once an ant finds a food source, it exploits the same and returns to the nest leaving pheromones on the path. The concentration of this pheromone can guide other ants in searching for food. When other ants find the pheromones they follow the path with a probability proportional to the concentration of the pheromone. Now if other ants also able to find the food source, they also leave pheromones during their return to the nest. As more ants find the path, the concentration of pheromone gets stronger. The pheromone evaporates with time and hence the longer paths will have more evaporation as compared to shorter paths. Ant colony optimization has been extensively used in solving many discrete optimization problems. Traveling salesman problem, robot path planning, minimum spanning tree, data mining, classification, scheduling problems are few examples where ACO has shown its strength in providing efficient solutions.

2   Evolutionary Computation

Creatures such as plants, animals, birds etc. have evolved and have been evolving continuously, by adapting themselves to the dynamic environment. The candidates who are strong enough (more fit to the environment) compared to others are likely to produce offspring(s) that are likely to survive. Darwinian evolution and laws of natural selection are responsible for this evolution. Evolutionary Computing draws its inspiration from biological evolution. It is hard to specify exactly when the first use of evolutionary principles was made to solve computational problems. However, to give some idea, we quote here De Jong et al. [6] "One of the first descriptions of the use of an evolutionary process for computer problem solving appeared in the articles by Friedberg [12] and Friedberg et al. [11]. This represented some of the early work in machine learning and described the use of an evolutionary algorithm for automatic programming, i.e. the task of finding a program that calculates a given inputoutput function. Other founders in the field remember a paper of Fraser [10] that influenced their early work, and there may be many more such forerunners depending on whom one asks." For more on the history of EC, readers are directed to De Jong et al. [6].

Fig. 1 How ants find the
shortest path to their next

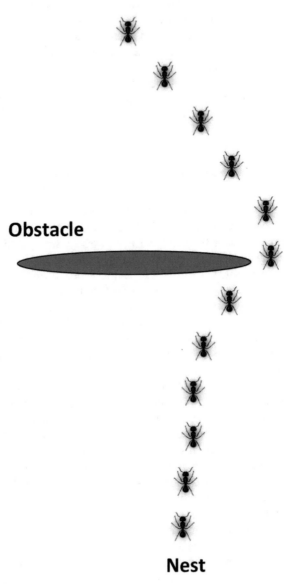

Fig. 1 How ants find the shortest path to their next

Evolutionary Computation (EC) primarily is used for solving optimization problems. Evolutionary Computation is the collective name for a range of problem-solving
techniques based on principles of biological evolution, such as natural selection and
genetic inheritance. These algorithms try to find globally optimal solutions. In a
very short span of time, these techniques have shown their applications to many

problems in diverse areas including engineering, science and agriculture. EC emulates the process of natural selection in the search procedure. The algorithms in the class of evolutionary computation begin by randomly generating a set (population) of potential solutions. Then a new population is obtained by iteratively modifying these potential solutions. The modification is done by iterative application of selection, crossover, and mutation operators. This process stochastically discards poor solutions and evolves more fit (better) solutions. Due to the very nature of these operators, it is expected that the evolved solutions will become better generation by generation (iteration).

2.1 Members of Evolutionary Computation

There are many families of algorithms that come under the umbrella of EC such as genetic algorithms (GAs) [13], Genetic Programming (GP) [2], Evolutionary Programming (EP) [9], and Evolutionary Strategy (ES) [4]. Except GP, the other members of EC solve optimization problems. On the other hand, GP generally finds programs that can solve a given problem. Genetic algorithms evolve based on Darwinian principle of survival of the fittest and encoding of individuals is usually done as binary vectors while, as mentioned earlier, GP although uses the same principle of survival of the fittest as GA, evolves individuals, that are programs. Evolutionary programming is inspired by the theory of evolution by means of natural selection; on the other hand, ES is a search technique based on the idea of adaptation and evolution, where encoding of individuals is done as a vector of reals. Differential evolution, on the other hand, differs from GA in the reproduction mechanism. While DE shares many similarities with other evolutionary algorithms, it differs significantly in the sense that in DE distance and direction information from the current population is used to guide the search process. Mutation is applied first to generate a trial vector, which is then used within the crossover operator to produce one offspring, while in a general EA, the crossover operator is applied first and then the mutation operator. Also DE mutation step sizes are influenced by differences between individuals of the current population while EA mutations are sampled from some probability distribution.

Some very distinguished characteristics of EC are: they can solve very unstructured problems, we just need to have a way to assess the quality (fitness) of a solution and they do not demand differentiability of the objective functions. Typically, evolutionary algorithms are criticized to be computationally very expensive and not suitable for solving really large scale problems. Although this is true in general, a recent breakthrough gives a great hope. A recent development reveals that a customized evolutionary algorithm can deal with integer linear programming problems involving billion variables [7].

3   Discussion

The philosophy behind these two approaches, SI and EC, is mainly rooted in the biological behaviors of natural objects. Both these approaches are inspired by natural phenomena. Both are search strategies to find an optimal solution. Swarm Intelligence is a study of the collective behavior emerged from social insects or animals working under very few rules. While evolutionary algorithms are adaptive heuristic search algorithms based on the evolutionary ideas of natural selection and genetics.

In Swarm Intelligence, individual members of the population have an identity, which it retains over time, in the form of temporarily linked movements. But in the evolutionary algorithm, population members die to be replaced by offspring. SI algorithms are inspired from food foraging behavior of swarms like birds, fishes, and ants, where the population members update their positions only in order to make them fit with respect to the environment. On the other hand, evolutionary algorithms are based on the Darwinian principle of survival of the fittest. In these algorithms, the whole population is replaced by new generation using the natural operators like crossover and mutation.

We closed our discussion in the previous section on Evolutionary Computing with an application of EC that solves an integer linear programming problem involving billion variables this may be the largest real-world constrained optimization problem that has ever been dealt with any optimization method. In [7] authors proposed a very fast population based method to obtain near optimal solutions. And this success story certainly emphasizes that possibly EC has not been explored adequately and in the right manner, and it is powerful enough to solve truly large problems. So there may be (are) more to get from it and hence there is a need for more research. It is important to study such algorithms because with the development of advanced technology, more data of varied nature are being generated and researchers/practitioners try to solve more complex large problems demanding high computation. For such problems, sometimes traditional algorithms of optimization become inappropriate. Moreover, complex systems are modeled with complicated multi-dimensional functions in most of the optimization problems and sometimes they cannot be amenable to classical optimization techniques. It is well known that classical or traditional optimization techniques have some limitations in solving real world optimization problems. This is mainly due to the inherent solution mechanisms of these techniques. Their efficiency is also very much dependent on the dimension of the problem as well as the structure of solution search space like convex or non-convex. For example, the simplex method can be applied only to solve linear objective function having linear constraints. But most of the optimization problems in the present scenario involve a variety of decision variables and complex structured objective and constraints (linear and/or nonlinear). Therefore, traditional or mathematical optimization procedures are generally not adequate or easy to use for their solutions. If we model such problems to make them solvable by classical methods then almost surely we will have to compromise with the exact formulation and hence compromise with the solution quality. Due to limitations of classical optimization algorithms in solving large-scale complex optimization problems, swarm intelligence and evolutionary computation based algorithms have become popular. These techniques are efficient and flexible. They can be modified and/or adapted to suit specific problem requirements. These nature inspired schemes take full advantage of modern computer systems. We emphasize that it is not our intention to give an impression that EC is an all-cure solution and always better than classical method. EC-based methods should be used only when a situation demands that. If a problem can be efficiently solved by a classical optimization method, then certainly that should be used. For example, if one needs to solve a quadratic optimization problem with linear constraints or a linear programming problem of reasonable size, one must use classical methods because such methods are backed up by good mathematical properties and convergence results.

4 Concluding Remarks

The EC and SI algorithms are better choices for some complex optimization problems because they do not demand much in terms of mathematical properties of objectives and constraints like convexity, continuity or explicit definition. These methods use a stochastic approach and they can be applied to a wider set of problems. But, this applicability comes with a cost, probabilistic convergence to the global optima. Also sometimes these algorithms are unable to manage a proper balance between exploration and exploitation of search space in their current states. Often the choice of parameters to guide the search process becomes difficult and the results may significantly depend on such choices. Considering these limitations, there are ample scopes of improvement in swarm intelligence and evolutionary algorithms. Modifications and developments is a continuous process in these fields to make these algorithms efficient, accurate and reliable.

Availability of high computational capacity inspired people to exploit EC and SI in solving complex problems more efficiently. Swarm and Evolutionary algorithms can provide tools to solve such problems. The No free lunch theorem, No single algorithm can be designated as the best algorithm provided it is tested over a sufficient number of problems always inspires researchers to develop new computationally intelligent algorithms. The swarm and evolutionary algorithms have also been extensively hybridized with other machine learning approaches. Recently deep learning with swarm and/or evolutionary algorithms has shown its potential as a very promising machine learning algorithm [13]. We note here that EC and SI enjoy benefits when the objective function is complex, non-differentiable, non-convex and so on. However, there are data mining problems, where in addition to complex formulation of the problem, the amount of data involved may be very large (for example, in atmospheric science, health care, astrophysics, and social media). In this era of big data, use of EC and SI to mine such data sets may pose a huge challenge to the researchers. This certainly does not mean that there is no scope for improvement, rather it suggests that we need more research effort in these areas.

References

1. Bansal, J.C., Sharma, H., Jadon, S.S., Clerc, M.: Spider monkey optimization algorithm for numerical optimization. Memetic Comput. **6**(1), 31–47 (2014)
2. Banzhaf, W., Nordin, P., Keller, R.E., Francone, F.D.: Genetic Programming: An Introduction, vol. 1. Morgan Kaufmann, San Francisco (1998)
3. Beni, G., Wang, J.: Swarm intelligence in cellular robotic systems. In: *Robots and Biological Systems: Towards a New Bionics?*, pp. 703–712. Springer, Berlin (1993)
4. Beyer, Hans-Georg, Schwefel, Hans-Paul: Evolution strategies—a comprehensive introduction. Nat. Comput. **1**(1), 3–52 (2002)
5. Bonabeau, E., Dorigo, M., Theraulaz, G: *Swarm Intelligence: From Natural to Artificial Systems*, Number 1. Oxford University Press, Oxford (1999)
6. De Jong, K., Fogel, D., Schwefel, H.-P.: *Handbook of Evolutionary Computation*, Chapter A history of evolutionary computation, pp. A2.3:1–12. CRC Press (1997)
7. Deb, K., Myburgh, C.: Breaking the billion-variable barrier in real-world optimization using a customized evolutionary algorithm. In *Proceedings of the 2016 on Genetic and Evolutionary Computation Conference*, pp. 653–660. ACM (2016)
8. Dorigo, M.: Optimization, learning and natural algorithms. Ph.D. Thesis, Politecnico di Milano, Italy (1992)
9. Fogel, L.J., Owens, A.J., Walsh, M.J.: Artifical Intelligence Through Simulated Evolution, vol. 1. Wiley, Hoboken (1967)
10. Fraser, A.S.: Simulation of genetic systems by automatic digital computers I. Introduction. Aust. J. Biol. Sci. **10**(4), 484–491 (1957)
11. Friedberg, R.M., Dunham, B., North, J.H.: A learning machine: part II. IBM J. Res. Dev. **3**(3), 282–287 (1959)
12. Friedberg, R.M.: A learning machine: Part I. IBM J. Res. Dev. **2**(1), 2–13 (1958)
13. Goldberg, D.E.. Optimization & machine learning. Genetic Algorithm in Search (1989)
14. Karaboga, D.: An idea based on honey bee swarm for numerical optimization. Technical report, Technical report-tr06, Erciyes University, Engineering Faculty, Computer Engineering Department (2005)
15. Kennedy, J., Eberhart, R.: Particle swarm optimization. In: 1995 IEEE International Conference on Neural Networks Proceedings (1942)
16. Passino, K.M.: Biomimicry of bacterial foraging for distributed optimization and control. IEEE Control Syst. **22**(3), 52–67 (2002)
17. Reynolds, C.W.: Flocks, herds and schools: a distributed behavioral model. In: ACM SIGGRAPH computer graphics, vol. 21, issue 4, pp. 25–34 (1987)
18. Yang, X.-S. (2009) Firefly algorithms for multimodal optimization. In *International Symposium on Sstochastic Algorithms*, pp. 169–178. Springer, Berlin (2009)

Particle Swarm Optimization

Jagdish Chand Bansal

Abstract Particle Swarm Optimization (PSO) is a swarm intelligence based numerical optimization algorithm, introduced in 1995 by James Kennedy, a social psychologist, and Russell Eberhart, an electrical engineer. PSO has been improved in many ways since its inception. This chapter provides an introduction to the basic particle swarm optimization algorithm. For better understanding of the algorithm, a worked-out example has also been given.

Keywords Particle swarm optimization · Swarm intelligence · Numerical optimization

1 Particle Swarm Optimization

Particle Swarm Optimization (PSO) is a swarm intelligent algorithm, inspired from birds' flocking or fish schooling for the solution of nonlinear, nonconvex or combinatorial optimization problems that arise in many science and engineering domains. In [4] Feng et al. used PSO to optimize parameters in radical basis function (RBF) network. In [2] PSO is modified to solve the two-sided U-type assembly line balancing (TUALB) problem. The results obtained by proposed approach showed that designing a two-sided assembly line in U-shaped layout provides shorter lines. Mousavi et al. applied PSO to optimize a supply chain network for a seasonal multi-product inventory system with multiple buyers, multiple vendors and warehouses with limited capacity owned by the vendors [8]. Apart from continuous version of PSO, binary PSO has also been applied extensively. Bansal et al. [1] proposed a modified binary PSO and applied it to solve various forms of knapsack problems. Jain et al. [5] proposed an improved-Binary Particle Swarm Optimization (iBPSO) for cancer classification. The next subsection presents the motivation and general framework of PSO procedure.

J. C. Bansal (✉)
Department of Mathematics, South Asian University, New Delhi, India
e-mail: jcbansal@gmail.com

© Springer International Publishing AG, part of Springer Nature 2019
J. C. Bansal et al. (eds.), *Evolutionary and Swarm Intelligence Algorithms*, Studies in Computational Intelligence 779,
https://doi.org/10.1007/978-3-319-91341-4_2

Russel Eberhart and James Kennedy

1.1 Motivation

Many bird species are social and form flocks for various reasons. Flocks may be of different sizes, occur in different seasons and may even be composed of different species that can work well together in a group. More eyes and ears mean increased opportunities to find food and improved chances of detecting a predator in time. Flocks are always beneficial for survival of its members in many ways [10]:

Foraging: A socio-biologist E. O. Wilson said that, "In theory at least, individual members of the school (swarm) can profit from the discoveries and previous experience of all other members of the school during the search for food" [11]. If for a group of birds, the food source is the same then some species of birds form flock in a non-competing way. In this way, more birds take advantage of discoveries of other birds about the location of the food.

Protection against Predator: A flock of birds have number of advantages in protecting themselves from the predator:

- More ears and eyes means more chances of spotting a predator or any other potential threat.
- A group of birds may be able to confuse or overwhelm a predator through mobbing or agile flights.
- In case of a group, large availability of prays reduces the danger for any single bird.

Aerodynamics: When birds fly in flocks, they often arrange themselves in specific shapes or formations. Those formations take advantage of the changing wind patterns based on the number of birds in the flock and how each bird's wings create different currents. This allows flying birds to use the surrounding air in the most energy efficient way.

However, the development of PSO requires simulation of some advantages of birds' flock, in order to understand an important property of swarm intelligence and therefore of PSO, it is worth mentioning some disadvantages of the birds' flocking. When birds form flock they also create some risk for them. More ears and more eyes means more wings and more mouths which result more noise and motion. In this situation, more predators can locate the flock causing a constant threat to the birds. A larger flock will also require a greater amount of food which causes more competition for food. This may result in death of some weaker birds of the group. It is important to mention here that PSO does not simulate the disadvantages of the birds' flocking behavior and therefore, during the search process killing of any individual is not allowed as in Genetic Algorithms where some weaker individuals die out. In PSO, all individuals remain alive and try to make themselves stronger throughout the search process. The improvement in potential solutions in PSO is due to cooperation while in evolutionary algorithms it is due to competition. This concept makes swarm intelligence different from evolutionary algorithms. In short, in evolutionary algorithms a new population is evolved in every generation / iteration while in swarm intelligent algorithms in every generation / iteration individuals make themselves better. Identity of the individual does not change over the iterations.

Mataric [7] gave the following rules for birds' flocking:

1. **Safe Wandering**: When birds fly they are not allowed to collide with each other and with obstacles.
2. **Dispersion**: Each bird will maintain a minimum distance with any other.
3. **Aggregation**: Each bird will also maintain a maximum distance with any other.
4. **Homing**: All birds will have potential to find a food source or the nest.

All these four rules have not been adopted in the simulation of birds flocking behavior while designing the PSO. In basic PSO model developed by Kennedy and Eberhart, safe wandering and dispersion rules are not followed for the movement of agents. In other words, during the movement in basic PSO agents are allowed to come closer to each other as they can. While aggregation and homing are valid in the PSO model. In PSO, agents have to fly within a particular region so that they can maintain a maximum distance with any other agent. This is equivalent to the fact that throughout the process, search remains within or at the boundaries of the search space. The fourth rule, homing says that any agent in the group may reach to the global optima.

For the development of PSO model, Kennedy and Eberhart followed five fundamental principles which determine whether a group of agents is a swarm or not [12]:

1. Proximity Principle: the population should be able to carry out simple space and time computations.

2. Quality Principle: the population should be able to respond to quality factors in the environment.
3. Diverse Response Principle: the population should not commit its activity along excessively narrow channels.
4. Stability Principle: the population should not change its mode of behaviour every time the environment changes.
5. Adaptability Principle: the population should be able to change its behaviour mode when it is worth the computational price.

1.2 Particle Swarm Optimization Process

Considering these five principles Kennedy and Eberhart developed a PSO model for function optimization. In PSO, the solution is obtained through a random search equipped with swarm intelligence. In other words, PSO is a swarm intelligent search algorithm. This search is done by a set of randomly generated potential solutions. This collection of potential solutions is known as *swarm* and each individual potential solution is known as a *particle*.

In PSO, the search is influenced by two types of learning by the particles. Each particle learns from other particles and it also learns from its own experience during the movement. The learning from others may be referred as *social learning* while the learning from own experience as *cognitive learning*. As a result from social learning, the particle stores in its memory the best solution visited by any particle of the swarm which we call as *gbest*. As a result of cognitive learning, the particle stores in its memory the best solution visited so far by itself, called *pbest*.

Change of the direction and the magnitude in any particle is decided by a factor called *velocity*. This is the rate of change in the position with respect to the time. With reference to the PSO, time is the iteration. In this way, for PSO, the velocity may be defined as the rate of change in the position with respect to the iteration. Since iteration counter increases by unity, the dimension of the velocity v and the position x becomes the same.

For a D-dimensional search space, the ith particle of the swarm at time step t is represented by a D-dimensional vector, $x_i^t = (x_{i1}^t, x_{i2}^t, \ldots, x_{iD}^t)^T$. The velocity of this particle at time step t is represented by another D-dimensional vector $v_i^t = (v_{i1}^t, v_{i2}^t, \ldots, v_{iD}^t)^T$. The previously best visited position of the ith particle at time step t is denoted as $p_i^t = (p_{i1}^t, p_{i2}^t, \ldots, p_{iD}^t)^T$. 'g' is the index of the best particle in the swarm. The velocity of the ith particle is updated using the velocity update equation in (1).

Velocity Update Equation:

$$v_{id}^{t+1} = v_{id}^t + c_1 r_1 (p_{id}^t - x_{id}^t) + c_2 r_2 (p_{gd}^t - x_{id}^t) \tag{1}$$

The position is updated using position update equation in (2).

Position Update Equation:
$$x_{id}^{t+1} = x_{id}^t + v_{id}^{t+1} \tag{2}$$

where $d = 1, 2, \ldots, D$ represents the dimension and $i = 1, 2, \ldots, S$ represents the particle index. S is the size of the swarm and c_1 and c_2 are constants, called cognitive and social scaling parameters, respectively or simply acceleration coefficients. r_1, r_2 are random numbers in the range [0, 1] drawn from a uniform distribution. It appears from Eqs. (1) and (2) that every particle's each dimension is updated independently from the others. The only link between the dimensions of the problem space is introduced via the objective function, i.e., through the locations of the best positions found so far *gbest* and *pbest* [9]. Equations (1) and (2) define the basic version of PSO algorithm. An algorithmic approach of PSO procedure is given in Algorithm 1:

Create and Initialize a D-dimensional swarm, S and corresponding velocity vectors ;
for *t= 1 to the maximum bound on the number of iterations* **do**
$\quad$ **for** *i=1 to S* **do**
$\quad\quad$ **for** *d=1 to D* **do**
$\quad\quad\quad$ Apply the velocity update equation 1;
$\quad\quad\quad$ Apply position update equation 2;
$\quad\quad$ **end**
$\quad\quad$ Compute fitness of updated position;
$\quad\quad$ If needed, update historical information for pbest and gbest;
$\quad$ **end**
$\quad$ Terminate if gbest meets problem requirements;
end

Algorithm 1: Basic Particle Swarm Optimization

1.3 Understanding Update Equations

The right hand side in the velocity update Eq. (1), consists of three terms [3]:

1. The previous velocity v, which can be thought of as a momentum term and serves as a memory of the previous direction of movement. This term prevents the particle from drastically changing direction.
2. The second term is known as the cognitive or egoistic component. Due to this component, the current position of a is attracted towards its personal best position. In this way, throughout the search process, a particle remembers its best position and thus prohibits itself from wandering.
   Here, it should be noted that $(p_{id} - x_{id})$ (superscript t is dropped just for simplicity) is a vector whose direction is from x_{id} to p_{id} which results the attraction of current position towards the particle's best position. This order of x_{id} and p_{id}

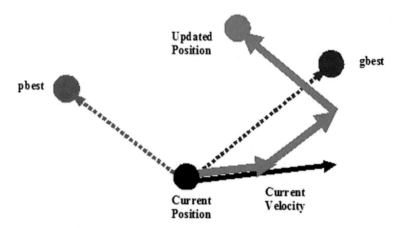

Fig. 1 Geometric Illustration of Particle's Movement in PSO Process

must be maintained for attraction of current position towards the particle's best position. If we write the second term using vector $(x_{id} - p_{id})$ then the current position will repel from the particle's best position.

3. The third term is called social component and is responsible for sharing information throughout the swarm. Because of this term a particle is attracted towards the best particle of the swarm, i.e. each particle learns from others in the swarm. Again the same reason stands here also to keep the order of x_{id} and p_{gd} in the vector $(p_{gd} - x_{id})$.

It is clear that cognitive scaling parameter c_1 regulates the maximum step size in the direction of the personal best position of that particle while social scaling parameter c_2 regulates the maximum step size in the direction of global best particle. Figure 1 presents a typical geometric illustration of a particle's movement in a 2-Dimensional space.

From the update equations, it is also clear that the PSO design of Kennedy and Eberhart follows five basic principals of PSO, described in Sect. 1.1. In the PSO process, calculations are carried out over a series of time steps in a D-dimensional space. Population at any time step, follows the direction guided by gbest and pbest, i.e. the population is responding to the quality factors and thus the quality principal is adhered to. Because of uniformly distributed random numbers r_1 and r_2 in the velocity update equation, a random allocation of current position between pbest and gbest justifies the diverse response principle. Principle of stability is also justified in PSO process because no particle of the swarm moves randomly but only when it receives a better information from gbest. The swarm changes when gbest changes and therefore adaptability principle is adhered to.

2 Particle Swarm Optimization Parameters

The convergence speed and the ability of finding optimal solution of any population based algorithm is greatly influenced by the choice of its parameters. Usually, a general recommendation for the setting of parameters of these algorithms is not possible as it is highly dependent upon the problem parameters. However, theoretical and/or experimental studies have been carried out to recommend the generic range for parameter values. Likewise other population based search algorithms, tuning of parameters for a generic version of PSO has always been a challenging task due the presence of stochastic factors r_1 and r_2 in the search procedure. The basic version of PSO enjoys the luxury of very few parameters. This chapter discusses parameters of only the basic version of PSO introduced in [6].

One radical parameter is the swarm size which is often set empirically on the basis of the number of decision variables in the problem and problem complexity. In general, 20–50 particles are recommended.

Another parameters are scaling factors, c_1 and c_2. As mentioned earlier, these parameters decide the step size of the particle for the next iteration. In other words, c_1 and c_2 determine the speed of particles. In basic version of PSO, $c_1 = c_2 = 2$ were chosen. With this choice, particle's speed increases without control which is good for faster convergence rate but harmful for better exploitation of the search space. If we set $c_1 = c_2 > 0$ then particles will attract towards the average of pbest and gbest. $c_1 > c_2$ setting will be beneficial for multimodal problems while $c_2 > c_1$ will be beneficial for unimodal problems. Small values of c_1 and c_2 will provide smooth particle trajectories during the search procedure while larger values of c_1 and c_2 will be responsible for abrupt movements with more acceleration. Adaptive acceleration coefficients have also been proposed by the researchers [13].

Stopping criterion is also a parameter not only for PSO but for any population based meta-heuristic algorithm. Popular stopping criteria are usually based on maximum number of function evaluations or iterations which are proportional to the time taken by the algorithm and acceptable error. A more efficient stoping criteria is based on the available search capacity of the algorithm. If an algorithm does not improve the solution with a significant amount upto a certain number of iterations, search should be stopped.

3 A Worked-Out Example

In this section, a numerical example is explained for better understanding of the working of PSO. For simplicity, following sphere function in two dimension is considered to minimize using PSO.

$$Minf(x_1, x_2) = x_1^2 + x_2^2; \text{ where } x_1, x_2 \in (-5, 5).$$

First, we generate swarm of size 5, randomly using uniform distribution in the range
$(-5, 5)$:
The position matrix x =

x_{ij}	1	2
1	4.7059	-0.7824
2	4.5717	4.1574
3	-0.1462	2.9221
4	3.0028	4.5949
5	-3.5811	1.5574

As mentioned in the PSO Algorithm 1, initialization of velocity vectors is also
required at this stage. Velocity corresponding to a particle is initialized in the range
$[-V_{max}, +V_{max}]$. Here V_{max}, the maximum velocity bound is a PSO parameter and
is usually set $V_{max} = X_{max}$.
    Therefore, in this example velocity vector is generated uniformly in the range
$[-5, 5]$.
The velocity matrix V =

v_{ij}	1	2
1	4.0579	-2.215
2	-3.7301	0.4688
3	4.1338	4.5751
4	1.3236	4.6489
5	-4.0246	-3.4239

Next step is the objective function evaluation for the current position matrix. The
fitness is usually the value of the objective function in the optimization problem
being solved. A solution with better objective function value represents a better fit
solution. Since the considered problem is a minimization problem, we will consider
a solution better fit if it has small objective function value. Substituting $x_{11} = 4.7059$
and $x_{12} = -0.7824$ in the objective function $f = x_1^2 + x_2^2$, we get 22.7576. Similarly,
calculating objective function value for other position vectors, we get the following
initial fitness matrix: It can be observed that the minimum of these 5 objective function
values corresponding to 5 particles is 8.5600. Therefore, the most fit solution of
this swarm is $x_3 = (-0.1462, 2.9221)$ which we call "*gbest*". Since this is the first
iteration, no previous iteration exists for the comparison and therefore every particle's
current position is also the "*pbest*" position.
    Now we proceed to the next iteration using PSO update equations. It should
be noted here that all calculations for velocity and position update are carried out
component wise.
    Let us consider to update the first particle $x_1 = (4.7059, -0.7824)$. First we will
update its first component $x_{11} = 4.7059$. The velocity component corresponding to

f_1	22.7576
f_2	38.1844
f_3	8.5600
f_4	30.1299
f_5	15.2497

x_{11} is $v_{11} = 4.0579$. Therefore, we will apply velocity update equation to v_{11} as follows: (considering $c_1 = c_2 = 2$ and $r_1 = 0.34, r_2 = 0.86$)

$$v_{11} = 4.0579 + 2 \times 0.34 \times (4.7059 - 4.7059) + 2 \times 0.86 \times (-0.1462 - 4.7059)$$
$$= -4.2877$$

Since the updated velocity component -4.2877 lies in the range $[-5, 5]$, we accept the value for updating position. In case of updated velocity component value goes beyond the pre-specified range, we will consider the nearest boundary value. For example, suppose the updated velocity component is -5.8345 then we will set it to -5 because the maximum bound of velocity on this side is -5. Now if the component is 6.8976, the updated velocity will be set equal to 5 because of the similar reason.

Now the position update equation for x_{11} is

$$x_{11} = 4.7059 + (-4.2877)$$
$$= 0.4182$$

Since the updated solution component lies in the search space $(-5, 5)$, we accept the solution. If the updated position does not lie within the given search space, there are many methods suggested by researchers to deal with the situation, some of them will be discussed in this chapter. For this example, we will randomly re-initialize the particle if the updated value falls outside the search space boundary.

Similarly, we will update the second component $x_{12} = -0.7824$. To update this, we will first apply velocity update equation on $v_{12} = -2.215$.

$$v_{12} = -2.215 + 2 \times 0.47 \times (-0.7824 - (-0.7824)) + 2 \times 0.91 \times (2.9221 - (-0.7824))$$
$$= 4.5272$$

Updated $v_{12} = 4.5272$ is in the range $[-5, 5]$ and therefore we will use this value to update x_{12}.

$$x_{12} = -0.7824 + 4.5272$$
$$= 3.7448$$

Updated x_{12} is again within the search space $(-5, 5)$ so we accept the solution.

Thus the first particle after applying the PSO update equations becomes:
$x_1 = (0.4182, 3.7448)$
We update all the particles using the same procedure.

Second Particle:

$$v_{21} = -3.7301 + 2 \times 0.34 \times (4.5717 - 4.5717) + 2 \times 0.86 \times (-0.1462 - 4.5717)$$
$$= -11.8449$$
$$x_{21} = 4.5717 + (-11.8449)$$
$$= -7.2732$$

Since the updated value of x_{21} is out of the search space, we re-initialize this x_{21} in the range $(-5, 5)$. Let $x_{21} = 3.4913$.

$$v_{22} = 0.4688 + 2 \times 0.12 \times (4.1574 - 4.1574) + 2 \times 0.06 \times (2.9221 - 4.1574)$$
$$= 0.3206$$
$$x_{22} = 4.1574 + 0.3206$$
$$= 4.4780$$

Thus the second particle after PSO updating becomes:
$x_2 = (3.4913, 4.4780)$

Third Particle:

$$v_{31} = 4.1338 + 2 \times 0.69 \times (-0.1462 - (-0.1462)) + 2 \times 0.34 \times (-0.1462 - (-0.1462))$$
$$= 4.1338$$
$$x_{31} = -0.1462 + 4.1338$$
$$= 3.9876$$
$$v_{32} = 4.5751 + 2 \times 0.69 \times (2.9221 - 2.9221) + 2 \times 0.34 \times (2.9221 - 2.9221)$$
$$= 4.5751$$
$$x_{32} = 2.9221 + 4.5751$$
$$= 7.4972 \text{ (exceeding the search space bounds.)}$$
$$x_{32} = \text{random value in the range } (-5, 5)$$
$$= 4.3399$$

Thus the updated third particle is $(3.9876, 4.3399)$.

Fourth Particle:

$$v_{41} = 1.3236 + 2 \times 0.18 \times (3.0028 - 3.0028) + 2 \times 0.23 \times (-0.1462 - 3.0028)$$
$$= -0.1249$$
$$x_{41} = 3.0028 + (-0.1249)$$
$$= 2.8779$$
$$v_{42} = 4.6489 + 2 \times 0.61 \times (4.5949 - 4.5949) + 2 \times 0.94 \times (2.9221 - 4.5949)$$
$$= 1.5040$$
$$x_{42} = 4.5949 + 1.5040$$
$$= 6.0989 \text{ (falls outside the search space bounds.)}$$
$$x_{42} = \text{random value in the range } (-5, 5)$$
$$= 2.5774$$

Updated fourth particle is (2.8779, 2.5774).

Fifth Particle:

$$v_{51} = -4.0246 + 2 \times 0.09 \times (-3.5811 - (-3.5811)) + 2 \times 0.39 \times (-0.1462 - (-3.5811))$$
$$= -1.3454$$
$$x_{51} = -3.5811 + (-1.3454)$$
$$= -4.9265$$
$$v_{52} = -3.4239 + 2 \times 0.65 \times (1.5574 - 1.5574) + 2 \times 0.10 \times (2.9221 - 1.5574)$$
$$= -3.1510$$
$$x_{52} = 1.5574 + (-4.0246)$$
$$= -2.4672$$

Updated fifth particle is (-4.9265, -2.4672).

Therefore, after this initial iteration, the updated velocity matrix v is shown in Table 1, the updated position matrix x and the fitness matrix are shown in Tables 2 and 3, respectively.

Table 1 Updated velocity matrix

v_{ij}	1	2
1	-4.2877	4.5272
2	-11.8449	0.3206
3	4.1338	4.5751
4	-0.1249	1.504
5	-1.3454	-3.151

Clearly, it can be seen that the minimum objective function value is 14.1984 which corresponds to the first particle. Therefore, *gbest* for the updated swarm is the first particle x_1.

Now we compare this *gbest* with the previous *gbest*, obviously updated gbest is not better than the previous one so for the carrying out the next iteration, we consider the gbest of previous iteration $(-0.1462, 2.9221)$.

Now for each particle, we observe the selection of *pbest*. It should be noted that *gbest* is for the whole swarm and *pbest* is for a particular particle.

For the first particle:

Fitness in the previous swarm = 22.7576

Fitness in the current swarm = 14.1984

Clearly, the fitness of current swarm is better than that of its previous, so we set $pbest_1 = (0.4182, 3.7448)$. On the other hand, if the fitness of current swarm would not be better than that of its previous then the current *pbest* and old *pbest* would be the same.

Similarly,

for the second particle: $pbest_2 = (3.4913, 4.4780)$;

for the third particle: pbest3 = (2.2534, 3.1379);

for the fourth particle: pbest4 = (1.6400, 1.3202) and

for the fifth particle: pbest5 = (2.2668, 2.0009).

The same procedure is continued until the termination criterion is attained.

As a final note to the chapter, "PSO is a dynamic population of active, interactive agents with no inherent intelligence". In PSO each individual teaches its neighbor, each individual learns from its neighbors. During the search procedure, potential solutions make better than random guesses using *Collaborative Trial and Error* strategies. These guesses are better than random search because they are informed by social learning. Since its inception, PSO has seen many changes which made

Table 2 Updated position matrix

x_{ij}	1	2
1	0.4182	3.7448
2	3.4913	4.4780
3	3.9876	4.3399
4	2.8779	2.5774
5	-4.9265	-2.4672

Table 3 Updated fitness values

f_1	14.1984
f_2	32.2416
f_3	34.7356
f_4	14.9252
f_5	30.3574

it a strong candidate for numerical optimization. Researchers have applied PSO to almost all kind of problems where a numerical optimization technique is expected to work.

Particle Swarm Optimization is quite flexible for modifications according to the problem requirements. Therefore, even after 23 years of its invention, there is enough scope to modify PSO and apply it to new complex optimization problems.

References

1. Bansal, J.C., Deep, K.: A modified binary particle swarm optimization for knapsack problems. Appl. Math. Comput. **218**(22), 11042–11061 (2012)
2. Delice, Y., Aydoğan, E.K., Özcan, U., İlkay, M.S.: Balancing two-sided u-type assembly lines using modified particle swarm optimization algorithm. 4OR **15**(1), 37–66 (2017)
3. Engelbrecht, A.P.: Computational Intelligence: An Introduction. Wiley.com (2007)
4. Feng, J., Tian, F., Jia, P., He, Q., Shen, Y., Fan, S.: Improving the performance of electronic nose for wound infection detection using orthogonal signal correction and particle swarm optimization. Sens. Rev. **34**(4), 389–395 (2014)
5. Indu, J., Jain, V.K., Jain, R.: Correlation feature selection based improved-binary particle swarm optimization for gene selection and cancer classification. Appl. Soft Comput. **62**, 203–215 (2018)
6. James, K., Russell, E.: Particle swarm optimization. In Proceedings of 1995 IEEE International Conference on Neural Networks, pp. 1942–1948 (1995)
7. Mataric, M.J.: Interaction and intelligent behavior. Technical report, DTIC Document (1994)
8. Mousavi, S.M., Bahreininejad, A., Nurmaya Musa, S., Yusof, F.: A modified particle swarm optimization for solving the integrated location and inventory control problems in a two-echelon supply chain network. J. Intell. Manuf. **28**(1), 191–206 (2017)
9. Trelea, I.O.: The particle swarm optimization algorithm: convergence analysis and parameter selection. Inf. Process. Lett. **85**(6), 317–325 (2003)
10. Webpage. http://birding.about.com/od/birdbehavior/a/why-birds-flock.htm
11. Wilson, E.: 0.(1975) Sociobiology: The New Synthesis (1980)
12. Yang, B.: Modified particle swarm optimizers and their application to robust design and structural optimization. Ph.D. thesis, Munchen, Technical University, Dissertation (2009)
13. Zhan, Z.-H., Xiao, J., Zhang, J., Chen, W.: Adaptive control of acceleration coefficients for particle swarm optimization based on clustering analysis. In: IEEE Congress on Evolutionary Computation, 2007. CEC 2007, pp. 3276–3282. IEEE (2007)

Artificial Bee Colony Algorithm Variants and Its Application to Colormap Quantization

Bahriye Akay and Kader Demir

Abstract This chapter presents the basis of Artificial Bee Colony (ABC) algorithm and the modifications that were incorporated to the algorithm to solve constrained, multi-objective and combinatorial type of optimization problems. In the modified ABC algorithm for constrained optimization, the greedy selection mechanism is replaced with Deb's rules to favor the search towards feasible regions. In the ABC algorithm proposed for multi-objective optimization, a non-dominated sorting procedure is employed to rank the individuals based on Pareto-dominance rules. Combinatorial type of problems can also be efficiently solved by the ABC algorithm incorporated with a local search compatible with combinatorial type problems. In the second part of the chapter, an application of the ABC algorithm to colormap quantization is presented. Results of the ABC algorithm was compared to those of k-means, fuzzy-c-means and particle swarm optimization algorithms. It can be reported that compared to the k-means and fuzzy-c-means algorithms, the ABC algorithm has the advantage of working with multi-criterion cost functions and being more efficient compared to particle swarm optimization algorithm.

Keywords Unconstrained ABC · Constrained ABC · Multi-objective ABC
Combinatorial ABC · ABC for color map quantization

1 Introduction

In nature, a swarm is a group of individual agents achieving a goal collectively. The goal may be protecting the swarm against predators, building a nest, retaining

B. Akay (✉)
Department of Computer Engineering, Erciyes University,
38039 Melikgazi, Kayseri, Turkey
e-mail: bahriye@erciyes.edu.tr

K. Demir
Institute of Natural and Applied Science, Erciyes University,
38039 Melikgazi, Kayseri, Turkey

© Springer International Publishing AG, part of Springer Nature 2019
J. C. Bansal et al. (eds.), *Evolutionary and Swarm Intelligence Algorithms*, Studies in Computational Intelligence 779,
https://doi.org/10.1007/978-3-319-91341-4_3

or breeding the population, exploiting the sources in the environment efficiently, etc. In a swarm, there is a task selection mechanism and division of labor to perform the goals, and the individuals are self-organized based on some local rules and interactions in a neighborhood. These low-level interactions lead to a global swarm behavior. Bonabeau et al. [6] defines the self-organization as a combination of positive feedback, negative feedback, fluctuations and multiple interactions. The positive feedback promotes the individuals to perform beneficial behaviors more frequently or to recruit the other individuals towards the convenient behaviors. Pheromone laying of ants or dancing of bees are examples to the positive feedback. The negative feedback mechanism abandonees unavailing patterns when the swarm goes to saturation due to the positive feedback effect. Evaporation of ant pheromone or abandonment of the exhausted sources by bees are examples of the negative feedback. The fluctuation brings creativity and innovation to explore new patterns. The multiple interaction is communication between neighbor agents of the swarm. The self-organization and the division of labor adapt the swarm to tolerate the external and internal changes. Swarm intelligence combining all the characteristics mentioned above has some advantages such as scalability, fault tolerance, adaptation, speed, modularity, autonomy, parallelism [18].

Ants, termites, bees, birds, and fishes live as swarms and perform some tasks collectively without a supervision. The collective and intelligent behaviors of these creatures lead some researchers to convey the collective intelligence to problem-solving techniques. Ant colony optimization by Dorigo [10] and Particle swarm optimization by Kennedy and Eberhart [19] are examples of swarm intelligence algorithms.

A bee swarm has many intelligent behavior patterns such as task division in the nest, mating, navigation, nest site selection, and foraging [16]. The foraging task is carried out by bees very efficiently by exhibiting all characteristics of self-organization and division of labor. The bees assigned to foraging task are divided into three categories: employed bees, onlooker bees and scout bees, which corresponds to the division of labor in the foraging task. The employed bees are responsible for exploiting the food sources and recruiting the other bees by dancing. The onlooker bees wait in the hive and choose a food source by watching the dances of the employed bees. The scout bees search for new unexplored sources. A food source exhausted by the exploitation is abandoned by its employed bee and the employed bee becomes a scout bee. Recruitment of the bees to profitable sources is a positive feedback phenomenon while the abandonment of the exhausted sources is a negative feedback phenomenon. Searching the undiscovered sources carried out by the scout bees is a fluctuation effect which brings innovation to available food sources. The bees' communication through the dancing includes information about the location and the quality of the sources, and this is the multiple interaction property of the self-organization.

Artificial bee colony (ABC) algorithm developed by Karaboga [13] is a swarm intelligence algorithm simulating the foraging behavior of honey bees. It is a successful tool for optimizing unconstrained and constrained, single-objective and multi-objective, and continuous and combinatorial design problems [16, 17]. This chapter gives a detailed explanation of ABC algorithm and its variants for solving unconstrained, constrained, multi-objective and combinatorial problems, and a section is devoted to its application to color map quantization.

2 ABC Algorithm

In 2005, Karaboga developed a new optimization algorithm simulating the foraging behavior of honey bees and called the new algorithm Artificial Bee Colony (ABC). The algorithm uses a population of food source positions. Each food source is an alternative solution to the optimization problem, and the nectar amount of a food source is the fitness of the solution. The algorithm tries to find an optimum solution (the most profitable source) by using some local and global search mechanisms in addition to various selection mechanisms performed by bees.

As mentioned in the previous section, bees are classified into three groups based on their food source selection type. These classes correspond to the phases of the algorithm. Main steps of the algorithm are given below:

1: Initialization
2: **repeat**
3:    Employed Bees' Phase
4:    Onlooker Bees' Phase
5:    Memorize the best solution achieved so far
6:    Scout Bee Phase
7: **until** Termination criteria is satisfied

Algorithm 1: Main steps of ABC algorithm

In the initialization phase of the algorithm, a food source population is generated randomly by using Eq. 1

$$x_{ij} = x_j^{min} + rand(0, 1)(x_j^{max} - x_j^{min}) \tag{1}$$

where $i = 1 \ldots SN$, $j = 1 \ldots D$, SN is the number of food sources, D is the number of design parameters, x_j^{min} and x_j^{max} are lower and upper boundary of jth dimension, respectively.

The initial population is improved through a foraging cycle of employed, onlooker and scout bees' phases. The foraging cycle is iterated until a termination criterion is satisfied. The termination criterion may be either reaching a maximum evaluation number or finding an acceptable function value.

In the employed bees' phase, the food source exploitation in real foraging behaviour is simulated by a local search in the neighborhood of the sources. The local search of the basic ABC algorithm is defined by Eq. 2:

$$v_{ij} = x_{ij} + \phi_{ij}(x_{ij} - x_{kj}) \tag{2}$$

where i is the current solution, k is a neighbor solution chosen randomly and ϕ_{ij} is a real random number within the range $[-1, 1]$ coming from uniform distribution. In the local search defined by Eq. 2, only one randomly chosen dimension of the current solution (parameter j) is changed. After the local search, a greedy selection between the current solution and its mutant is carried out to select the better one to survive. The better solution is retained in the population and the other one is discarded. The local search and the greedy selection are applied to each food source in the population.

Various modifications in the local search of ABC has been proposed in the literature [17].

Once the employed bees' phase is completed, onlooker bees' phase is performed. In the onlooker bees' phase, the neighborhood of the food sources is searched to find better solutions like in the employed bees' phase. Unlike the employed bees' phase, the search is not conducted in the vicinity of each solution one by one. Instead, the solutions that will be included in the search are selected stochastically depending on their fitness values, that is, high quality solutions are more likely to be selected. This is the positive feedback property of the ABC algorithm. Each solution is assigned a probability (Eq. 3) proportional to its fitness value.

$$p_i = \frac{fitness_i}{\sum_{i=1}^{SN} fitness_i} \tag{3}$$

After calculating the probability values, a fitness-based selection scheme is employed to give higher chance to better solutions. The selection scheme may be roulette wheel, ranking based, stochastic universal sampling, tournament selection, or another selection scheme. In the basic ABC, roulette wheel selection scheme is used. This is an analogy of a real honey bee colony in which better food sources attract the attention of more bees depending on the information taken from dancing employed bees. Once SN solutions are selected probabilistically and local searches are conducted in the vicinity of these solutions and then, the greedy search is applied to select better solutions as in the employed bee phase. In both the employed bees' and onlooker bees' phases, if a solution can not be improved by the local search, its counter is incremented by one. This counter holds how many times the solution is exploited and retained in the population. Hence, it is an analogy of the number of food source exploitations by bees in real life. In real life, nectar of a source is exhausted at the end of the exploitations as simulated in the ABC algorithm such that if a source is exploited sufficiently, this source is abandoned by its bee. The counters are used to determine exploitation sufficiency and exhaustion. If a counter exceeds a limit, the solution associated with the counter is assumed to be exhausted and is replaced with a new solution produced randomly by Eq. 1. Checking the counters and if required, producing a random solution operations comprise the scout bee phase.

Regarding all the phases, the algorithm has three control parameters: the number of food sources, the maximum number of cycles, and the limit used to determine the exhausted sources.

For real parameter optimization, ABC algorithm has been modified to increase its convergence ability [3]. In the basic ABC algorithm, only one dimension of the parameter vector is changed to produce a mutant solution. In the modified ABC algorithm proposed in [3], more than one dimension may be changed by Eq. 4.

$$v_{ij} = \begin{cases} x_{ij} + \phi_{ij}(x_{ij} - x_{kj}), & if \ R_{ij} < MR \\ x_{ij}, & \text{otherwise} \end{cases} \tag{4}$$

where R_{ij} is drawn from the standard uniform distribution on the interval $(0,1)$ and modification rate, MR, controls the frequency of perturbation. Therefore, for each dimension j, if $R_{ij} < MR$ is satisfied, v_{ij} is produced by the local search, otherwise v_{ij} equals x_{ij}.

In [3], another modification on the magnitude of ϕ_{ij} has been proposed. ϕ_{ij} is drawn at random within the range $[-1, 1]$ in the basic ABC, while in the modified ABC, it is drawn within the range bounded by a scale factor, $[-SF, SF]$. SF is scaling factor and it has been also proposed to change SF automatically based on Rechenberg's 1/5 rule for fine tuning purpose. If the ratio of successful mutations to all mutations is less than 1/5, SF is decreased, otherwise, SF is increased in order to speed up the search.

2.1 ABC Algorithm for Single-Objective Constrained Optimization

A constrained optimization problem (5) is defined as finding a parameter vector x that minimizes an objective function $f(x)$ subject to inequality and/or equality constraints:

$$\begin{aligned} & \text{minimize } f(x), \ x = (x_1, \ldots, x_n) \in \mathbb{R}^n \\ & \qquad\qquad l_i \le x_i \le u_i, \qquad\qquad i = 1, \ldots, n \\ & \text{subject to :} \quad g_j(x) \le 0, \qquad\qquad \text{for } j = 1, \ldots, q \\ & \qquad\qquad h_j(x) = 0, \qquad\qquad \text{for } j = q+1, \ldots, m \end{aligned} \tag{5}$$

The objective function f is defined on a search space, $\mathbb{S}$, which is defined as a n-dimensional rectangle in $\mathbb{R}^n$ ($\mathbb{S} \subseteq \mathbb{R}^n$). Domains of variables are defined by their lower and upper bounds. A feasible region $\mathbb{F} \subseteq \mathbb{S}$ is defined by a set of m additional constraints ($m > 0$) and x is defined on feasible space ($x \in \mathbb{F} \in \mathbb{S}$). At any point

$x \in \mathbb{F}$, constraints g_j that satisfy $g_j(x) = 0$ are called active constraints at x. By extension, equality constraints h_j are also called active at all points of $\mathbb{S}$ [24].

In order to cope with the constraints, basic optimization algorithms use an approach [20] such as transforming infeasible solutions to feasible ones with some operators [22, 25] or using penalty functions [5, 11, 12, 21] or making a clear distinction between feasible and infeasible solutions [7, 23, 28–30].

The ABC algorithm for constrained optimization [14] is integrated with Deb's rules [7] to make a distinction between feasible and infeasible solutions. The greedy selection mechanism can not be used here because infeasible solutions violating the constraints may have lower cost values.

- Any feasible solution is preferred to any infeasible solution ($violation_j > 0$) (solution i is dominant),
- Among two feasible solutions, the one having better objective function value is preferred ($f_i < f_j$, solution i is dominant),
- Among two infeasible solutions ($violation_i > 0$, $violation_j > 0$), the one having smaller constraint violation is preferred ($violation_i < violation_j$, solution i is dominant).

Due to Deb's rules, the ABC algorithm does not need to be started with an initial population of only feasible solutions. This means that the population is a mixture of feasible and infeasible solutions and the probability assignment scheme of ABC should also be changed in order to assign probability values to infeasible solutions. This leads to a modification on the probability calculation equation (Eq. 3). The modified equation is given by Eq. 6:

$$
p_i = \begin{cases} 0.5 + \left(\frac{fitness_i}{\sum_{j=1}^{sn} fitness_j} \right) * 0.5 & \text{if solution is feasible} \\ \left(1 - \frac{violation_i}{\sum_{j=1}^{sn} violation_j} \right) * 0.5 & \text{if solution is infeasible} \end{cases} \tag{6}
$$

where $violation_i$ is sum of the constraint values outside the feasible region. The Eq. 6 assigns probability within the range $[0, 0.5]$ to infeasible solutions proportional to their violation amounts, and within the $[0.5, 1]$ to feasible solutions depending on their fitness values.

Another modification is introduced in the neighborhood production. Since the basic ABC algorithm only changes one parameter of the parameter vector to produce a mutant solution, Eq. 4 can again be used to improve the convergence rate.

ABC algorithm for constrained optimization [14] checks the exhausted sources at some periods instead of at each cycle. This introduces a new control parameter which determines the periods, called scout production period (SPP). At each SPP period, it is controlled whether there is a food source exceeding limit or not.

2.2 ABC Algorithm for Multi-objective Optimization

When a problem has more than one objective function, it is called multi-objective optimization problem (MOP), and MOPs have more than one optimal solution known as Pareto-optimal solutions (PS) or non-dominated solutions. A MOP can be defined as follows:

$$\min/\max F(x) = (f_1(x), \ldots, f_m(x))^T$$
$$\text{subject to } x \in \Omega \qquad (7)$$

where Ω is decision variable space, $F : \Omega \rightarrow R^m$ is objective vector and R^m is the objective space.

As in the constrained optimization, in multi-objective optimization the basic algorithms are extended using different selection mechanisms in order to handle generally conflicting objectives. Omkar et al. presented a vector evaluated multi-objective ABC (MOABC) algorithm which uses sub-populations for each objective function and shuffles them [26]. Akay proposed two versions of multi-objective ABC algorithm. One of them is asynchronous MOABC which uses Pareto-dominance-based selection scheme (A-MOABC/PD) and the other one is synchronous MOABC which uses non-dominated sorting (S-MOABC/NS) [1]. A-MOABC/PD uses Pareto-dominance rules instead of greedy selection and non-dominated solutions generated form the Pareto-front solutions. In Pareto-dominance, a solution vector x is partially less than another solution y $(x \prec y)$, when none of the objectives of y is less than those of x, and at least one objective of y is strictly greater than that of x. If x is partially less than y, the solution x dominates y or y is inferior to x. Any solution which is not dominated by other solution is said to be non-dominated or non-inferior [32].

Another difference between the basic ABC algorithm and A-MOABC/PD is related with the cost function evaluation which considers Pareto rank, distance and Gibbs distribution probability of a solution. Each solution is assigned a cost value using Eq. 8:

$$f_i = R(i) - TS(i) - d(i) \qquad (8)$$

where $R(i)$ is the Pareto rank value of the individual i, $S(i) = -p_T(i)logp_T(i)$, where $T > 0$ is temperature. $p_T(i) = (1/Z)exp(-R(i)/T)$ is the Gibbs distribution, $Z = \sum_{i=1}^{N} exp(-R(i)/T)$ is called the partition function, and N is the population size [37]. $d(i)$ is the crowding distance calculated by a density estimation technique [9]. For minimization problems, a fitness value is assigned to each solution inversely proportional to the function value (Eq. 8) and onlookers are distributed to the sources using the probability values calculated by Eq. 3 as in the basic ABC algorithm.

In the second version of MOABC (S-MOABC/NS) proposed by Akay [1], non-dominated sorting, which uses a ranking selection method and a niche method, is applied to search Pareto-optimal solutions. Fast non-dominated sorting method can be found in [9]. Non-dominated sorting procedure can be applied asynchronously or synchronously in ABC. When it is used asynchronously, the population is sorted after each mutant solution is produced and evaluated. When it is used synchronously, the population is sorted after each phase of the algorithm.

Steps of S-MOABC/NS are given in Algorithm 2.

Data: Set the control parameters of the ABC algorithm.
CS: Number of Foods,
MCN: Maximum Cycle Number,
$limit$: Maximum number of trial for abandoning a source,
M: Number of Objectives
begin
    //Initialization;
    for $s = 1$ *to* CS **do**
        $X(s) \longleftarrow$ random solution by Eq. 1; $f_{si} \longleftarrow f_i(X(s)), i = 1 \ldots M$; $trial(s) \longleftarrow \emptyset$;
    end
    $cycle = 1$;
    while $cycle < MCN$ **do**
        //Employed Bees' Phase;
        for $s = 1$ *to* CS **do**
            $x' \longleftarrow$ a new solution produced by Eq. 2;
            $f_i(x') \longleftarrow$ evaluate objectives of new solution, $i = 1 \ldots M$;
            $X(CS + s) = x'$
        end
        $\mathcal{F} \longleftarrow$ Non-dominated sorting(X) ;
        Select best CS solutions based on rank and crowding distance to form new population ;
        Assign fitness values to solutions depending on the cost values defined by Eq. 8;
        Calculate probabilities for onlookers by (3);
        //Onlooker Bees' Phase;
        $s \longleftarrow 0$;
        $t \longleftarrow 0$;
        while $t < CS$ **do**
            $r \longleftarrow rand(0, 1)$;
            //Stochastic sampling;
            if $r < p(s)$ **then**
                $t \longleftarrow t + 1$;
                $x' \longleftarrow$ a new solution produced by Eq. 2;
                $f_i(x') \longleftarrow$ evaluate objectives of new solution, $i = 1 \ldots M$;
                $X(CS + s) = x'$
            end
            $s \longleftarrow (s + 1) \ mod \ (CS - 1)$;
        end
        $\mathcal{F} \longleftarrow$ Non-dominated sorting(X) ;
        //Scout bee phase;
        $mi = \{s : trial(s) = max(trial) \wedge X(s) \notin \mathcal{F}\}$;
        if $trial(mi) > limit$ **then**
            $X(mi) \longleftarrow$ random solution by Eq. 1;
            $f_{i,mi} = f_i(X(mi))$;
            $trial(mi) \longleftarrow \emptyset$;
        end
        $cycle + +$;
    end
end

Algorithm 2: S-MOABC/NS Algorithm

In order to observe the convergence process of an multi-objective algorithm quantitatively, some performance metrics are used such as inverted generational distance (IGD) [36], hypervolume [33, 35], and spread [8]. Synchronous MOABC based on non-dominated sorting procedure is shown to approximate to well-distributed and high quality non-dominated fronts [1].

2.3   ABC Algorithm for Combinatorial Optimization

Combinatorial optimization finds the minimum cost set, I, among all subsets of the power set of a finite set E [2]. Therefore, the algorithms tackling with combinatorial optimization problems should search the space of the finite set E to find the global optimum. Algorithms are required to have efficient search operators to reach the global optimum in polynomial times. However, there are some problems that can not be solved in polynomial times, called NP-hard problems, and the heuristic algorithms try to find good approximate solutions in polynomial times. ABC algorithm has been used to solve combinatorial problems such as leaf-constrained minimum spanning tree [31], TSP [2, 15], lot-streaming flow shop scheduling problem [27].

In [2], 2-opt local search operator is used in neighbor solution production. In 2-opt local move, a different edge is searched between current solution and a randomly chosen neighbor, and then the different edge is constructed on the current solution by re-linking ith and jth cities by 2-opt move (Fig. 1).

In [15], a mutation operator called Greedy Sub Tour Mutation (GSTM) proposed by Albayrak and Allahverdi is employed [4] and tested on TSP problems.

GSTM operator that is developed to solve travelling salesman problem by genetic algorithm which combines classical and greedy techniques to avoid getting stuck a local minimum. The GSTM operator reaches local solutions faster by applying greedy search methods and random distortion on these solutions like classical mutation operators.

Main steps of GSTM operator are given below.

1. Select solution x_k randomly in the population. ($k \neq i$).
2. Select a city x_{ij} randomly.
3. Select the value of searching way parameter $\phi \in \{-1, 1\}$ randomly.
4. if ($\phi=1$) then

   – The city visited before x_{kj} is set as the previous city of x_{ij}.

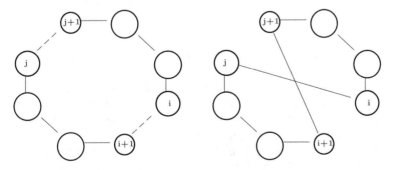

Fig. 1 2-opt move

5. else

   - The city visited after x_{kj} is set as the next city of x_{ij}.

6. endif
7. A new closed tour $\hat{T}$ is generated.
8. By this new connection, an open sub tour T^* is generated. R_1: first city of T^* and R_2: last city of T^*
9. if (random $< P_{RC}$) then

   - Subtract $[R_1 - R_2]$ tour from tour T ($T^\# \leftarrow T - [R_1 - R_2]$; $T^* \leftarrow [R_1 - R_2]$)
   - Hold on T^* sub-tour $T^\#$ so that there will be minimum extension.

10. else

   - if (random $< P_{CP}$) then
     - Copy $[R_1 - R_2]$ sub-tour in the T tour, $T^* \leftarrow [R_1 - R_2]$
     - Add each city of T^* to the T starting from the position R_1 by rolling or mixing with probability P_L.
   - else
     - Select randomly one neighbor from neighbor lists for the points R_1 and R_2 (NL_{R_1} and NL_{R_2}).
     - Invert the points NL_{R_1} or NL_{R_2} that provides maximum gain such a way that these points will be neighbors to the points R_1 or R_2.
     - Repeat if the inversion is not taken place.
   - endif

11. endif

Here, P_{CP} is correction and perturbation probability, P_{RC} is reconnection probability and P_L is linearity probability. These are parameters in the GSTM operator.

3   An Application of ABC Algorithm to Colormap Quantization

An image is a matrix of pixels and each pixel in a RGB image is represented by red, green and blue values coded by 8 bits, totally 24 bits and 16.7 million colors. As the number of colors used in an image increases, quality of the image and the similarity with the real scene also increase. However, this increases the image size at the same time and it is hard for hardware-restricted devices to display and to process high-resolution images. Hence, the images are required to be compressed to occupy less space and to be transferred easily. The purpose is representing the image using less bits and maximizing the similarity with the original image, which can be achieved by colormap quantization. Because human perception cannot sense 16.7 million different colors in an image [34], reducing the number of colors is reasonable.

Colormap quantization is composed of two steps. The first step is reducing the color palette depending on the desired number of colors and the second step is reconstructing the image based on the new color palette. The quality of the image is correlated with the colors in the palette. Therefore, the colors are chosen carefully for a high-quality image. Assigning the colors to the palette in order to maximize the similarity between the original image and the reduced image is an optimization problem.

In this section, the ABC algorithm has been applied to select the optimal color values to construct a color palette and its performance has been compared to those of k-means (KM), fuzzy-c-means(FCM), and particle swarm optimization (PSO) algorithms.

3.1 Experiments

In the experiments, three real images (Lena, Baboon and Peppers shown in Fig. 2a–c and two synthetic test images shown in Fig. 3a, b were used. The algorithms were implemented by using MATLAB environment.

(a) Lena Image   (b) Baboon Image   (c) Peppers Image

Fig. 2 Real test images used in the experiments

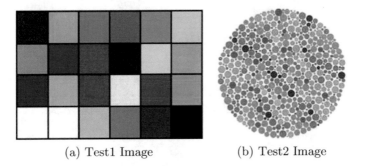

(a) Test1 Image   (b) Test2 Image

Fig. 3 Synthetic test images used in the experiments

The number of colors investigated in the experiments were 4, 8 and 16. Different control parameter configurations were investigated to find the best parameter sets, and the best parameter sets for each algorithm has been employed in the experiments. Runs with different control parameter values were repeated 108 times for ABC algorithm, 126 times for PSO algorithm, 27 times for KM and 27 times for FCM algorithms. Computing platform was 64 GB of Ram, Xeon E5-2640 CPU and 300 GB of disk. The maximum iteration number was 100 for k-means and FCM algorithms because they did not improve the results anymore after this iteration number. At the end of repeated experiments, values of the control parameters yielding the best performance were determined. Control parameters of the PSO algorithm were set to 20, 250, 0.5, 1.8, and 1.8 for swarm size, maximum generation number, inertia weight, cognitive component, and social component, respectively. Control parameters of the ABC algorithm were set to 20, 250, 0.8, 75 for colony size, maximum cycle number, modification rate, and limit, respectively.

All the experiments were repeated 30 times with different seed numbers using the best control parameter set and the results were presented in Tables 1, 2, 3, 4 and 5 for Lena, Peppers, Baboon, the first and the second test image, respectively. The cost function was the summation of the Euclidian distance of each pixel to the palette value and the mean square error of the quantized image and the original image.

Table 1 Results of the algorithms for the Lena image, NC: The number of Colors in the palette, Std. Dev.: Standard Deviation

NC = 4

	KM	FCM	PSO	ABC
Best	0.1040	0.1033	0.1031	0.1031
Mean	0.1040	0.1033	0.1187	0.1039
Median	0.1040	0.1033	0.1171	0.1038
Std. Dev.	0	0	0.0137	0.00074420

NC=8

	KM	FCM	PSO	ABC
Best	0.0726	0.0724	0.0732	0.0731
Mean	0.0730	0.0725	0.0960	0.0774
Median	0.0727	0.0725	0.0947	0.0768
Std. Dev.	0.00097693	0.000018097	0.0154	0.0031

NC = 16

	KM	FCM	PSO	ABC
Best	0.0519	0.0521	0.0675	0.0598
Mean	0.0525	0.0525	4.0867	0.0691
Median	0.0524	0.0524	0.0874	0.0681
Std. Dev.	0.00043861	0.00049088	8.1360	0.0060

Table 2 Results of the algorithms for the Pepper image, NC: The number of Colors in the palette, Std. Dev.: Standard Deviation

NC = 4

	KM	FCM	PSO	ABC
Best	0.1489	0.1475	0.1462	0.1462
Mean	0.1563	0.1475	0.1572	0.1469
Median	0.1489	0.1475	0.1470	0.1469
Std. Dev.	0.0135	0.00000095048	0.0168	0.00053633

NC = 8

	KM	FCM	PSO	ABC
Best	0.0950	0.0944	0.0952	0.0959
Mean	0.0969	0.0947	0.1167	0.1030
Median	0.0973	0.0946	0.1166	0.1009
Std. Dev.	0.0025	0.00038861	0.0106	0.0060

NC = 16

	KM	FCM	PSO	ABC
Best	0.0703	0.0708	0.0783	0.0794
Mean	0.0712	0.0710	0.0922	0.0898
Median	0.0711	0.0709	0.0908	0.0870
Std. Dev.	0.00065217	0.00019959	0.0089	0.0102

From the results of real images given in Tables 1, 2 and 3, when the number of colors was 4, the ABC and PSO algorithms were better than the KM and FCM algorithms. When the number of colors was 8, FCM seemed to be better and when the number of colors was 16, KM was better than the other algorithms. It can be said that when the number of colors was low and the image was more quantized, heuristic optimization algorithms were better than iterative KM and FCM algorithms. Between the PSO and ABC algorithms, the ABC algorithm was more stable based on the standard deviation metric.

When we analyze the results of synthetic test images which have strict transitions between the regions, the ABC algorithm produced better results in terms of cost function and visual results.

In Table 6, running times of the algorithms were presented. From the results, KM had the smallest running time, FCM was the first runner-up and ABC and PSO had similar running times.

These results show us that KM and FCM are fast algorithms. However, the PSO and ABC algorithms are flexible such that the cost function can be defined in a multi-objective approach which means the Euclidian distance of each pixel to palette value, the mean square error of the quantized image and the original image or structural similarity index measuring visual correlation can be defined separate cost functions to be optimized.

Table 3 Results of the algorithms for the Baboon image, NC: The number of Colors in the palette, Std. Dev.: Standard Deviation

NC = 4

	KM	FCM	PSO	ABC
Best	0.1771	0.1765	0.1760	0.1760
Mean	0.1771	0.1765	0.1792	0.1764
Median	0.1771	0.1765	0.1764	0.1763
Std. Dev.	0.0000041652	0.00000022982	0.0063	0.00031309

NC = 8

	KM	FCM	PSO	ABC
Best	0.1244	0.1242	0.1249	0.1244
Mean	0.1249	0.1244	0.1385	0.1308
Median	0.1247	0.1243	0.1394	0.1294
Std. Dev.	0.00046688	0.00018471	0.0083	0.0045

NC = 16

	KM	FCM	PSO	ABC
Best	0.0931	0.0942	0.0983	0.0976
Mean	0.0934	0.0945	0.1107	0.1034
Median	0.0934	0.0943	0.1115	0.1028
Std. Dev.	0.00038386	0.00067001	0.0071	0.0040

Table 4 Results of the algorithms for the test image 1, NC: The number of Colors in the palette, Std. Dev.: Standard Deviation

NC = 8

	KM	FCM	PSO	ABC
Best	0.1980	0.1931	0.1994	0.1878
Mean	0.2091	0.1946	0.2177	0.1983
Median	0.2053	0.1932	0.2127	0.1981
Std. Dev.	0.0133	0.0029	0.0201	0.0095

Table 5 Results of the algorithms for the test image 2, NC: The number of Colors in the palette, Std. Dev.: Standard Deviation

NC = 8

	KM	FCM	PSO	ABC
Best	0.1008	0.0968	0.0994	0.0902
Mean	0.1054	0.0968	0.1080	0.0923
Median	0.1032	0.0968	0.1080	0.0927
Std. Dev.	0.0063	0.0000018679	0.0074	0.0014

Table 6 Running times of algorithms (sc)

# of Colors	KM	FCM	ABC	PSO
4	0.26	0.49	3.88	3.87
8	0.31	0.83	4.05	4.03
16	0.43	1.54	4.37	4.32

4 Conclusion

Artificial Bee Colony algorithm is a powerful swarm intelligence-based algorithm which models the foraging behavior of honey bees. In this chapter, the ABC algorithm is presented and the schemes for different types of optimization problems are reviewed.

Basic ABC algorithm is developed to optimize unconstrained optimization problems. In order to solve the constrained optimization problems, the greedy selection scheme of the basic ABC algorithm is replaced with a selection scheme based on Deb's rules which favor the feasible region .

In order to be able to solve multi-objective optimization problems and to produce Pareto-solutions, non-dominated sorting mechanism is integrated to the basic algorithm and the selection is performed after each phase of the algorithm synchronously.

In addition to continuous optimization problems, combinatorial optimization problems can be solved by the ABC algorithm by a change in the neighbor solution production mechanism. The original neighbor solution mechanism can be replaced by a local search operator such as 2-opt move or GSTM operator.

The ABC algorithm has been applied to many different type of optimization problems in different fields including filter design, hydrology, civil engineering, mechanic, control systems, scheduling, data mining, circuit design etc.,. In the second part of the chapter, ABC algorithm is applied to color map quantization which represents an image by less bits. The ABC algorithm is used to select the optimal color values in the palette which maximize the similarity of the image with the original one. The performance of ABC is compared to those of k-means, fuzzy-c-means and particle swarm optimization algorithms. From the results, it can be said that the ABC algorithm can be used as a colormap quantization tool and has the flexibility to optimize more than one criteria compared to k-means and fuzzy-c-means algorithms.

Based on the studies on the ABC algorithm, the computational complexity of ABC algorithm is an important gap to fill in the existing literature as for the other computational intelligence algorithms.

References

1. Akay, B.: Synchronous and asynchronous pareto-based multi-objective artificial bee colony algorithms. J. Global Optim. **57**(2), 415–445 (2013)
2. Akay, B., Aydogan, E., Karacan, L.: 2-opt based artificial bee colony algorithm for solving traveling salesman problem pp. 666–667 (2011)

3. Akay, B., Karaboga, D.: A modified artificial bee colony algorithm for real-parameter optimization. Inf. Sci. **192**, 120–142 (2012)
4. Albayrak, M., Allahverdi, N.: Development a new mutation operator to solve the traveling salesman problem by aid of genetic algorithms. Expert Syst. Appl. **38**(3), 1313–1320 (2011)
5. Bean, J., Hadj-Alouane, A.B.: A Dual Genetic Algorithm for Bounded Integer Programs. Technical Report TR 92-53, Department of Industrial and Operations Engineering, The University of Michigan (1992), to appear in R.A.I.R.O.-R.O. (invited submission to special issue on GAs and OR)
6. Bonabeau, E., Dorigo, M., Theraulaz, G.: Swarm Intelligence: From Natural to Artificial Systems. Oxford University Press, Inc., New York, NY, USA (1999). http://portal.acm.org/citation.cfm?id=328320
7. Deb, K.: An efficient constraint handling method for genetic algorithms. Comput. Methods Appl. Mech. Eng. **186**(2–4), 311–338 (2000)
8. Deb, K., Agrawal, S., Pratap, A., Meyarivan, T.: A Fast Elitist Non-Dominated Sorting Genetic Algorithm for Multi-Objective Optimization: NSGA-II. KanGAL report 200001, Indian Institute of Technology, Kanpur, India (2000)
9. Deb, K., Pratap, A., Agarwal, S., Meyarivan, T.: A fast and elitist multiobjective genetic algorithm: NSGA-II. IEEE Trans. Evol. Comput. **6**(2), 182–197 (2002)
10. Dorigo, M., Maniezzo, V., Colorni, A.: Positive feedback as a search strategy. Technical Report 91-016, Politecnico di Milano, Italy (1991)
11. Homaifar, A., Lai, S.H.Y., Qi, X.: Constrained optimization via genetic algorithms. Simulation **62**(4), 242–254 (1994)
12. Joines, J., Houck, C.: On the use of non-stationary penalty functions to solve nonlinear constrained optimization problems with GAs. In: Fogel, D. (ed.) Proceedings of the First IEEE Conference on Evolutionary Computation. pp. 579–584. IEEE Press, Orlando, Florida (1994)
13. Karaboga, D.: An idea based on honey bee swarm for numerical optimization. Technical Report TR06, Erciyes University, Engineering Faculty, Computer Engineering Department (2005)
14. Karaboga, D., Akay, B.: A modified artificial bee colony (abc) algorithm for constrained optimization problems. Appl. Soft Comput. **11**(3), 3021–3031 (2011)
15. Karaboga, D., Gorkemli, B.: A combinatorial artificial bee colony algorithm for traveling salesman problem. In: 2011 International Symposium on Innovations in Intelligent Systems and Applications (INISTA), pp. 50–53 (2011)
16. Karaboga, D., Akay, B.: A survey: algorithms simulating bee swarm intelligence. Artif. Intell. Rev. **31**(1–4), 61–85 (2009)
17. Karaboga, D., Gorkemli, B., Ozturk, C., Karaboga, N.: A comprehensive survey: artificial bee colony (ABC) algorithm and applications. Artif. Intell. Rev. 1–37 (2012)
18. Kassabalidis, I., El-Sharkawi, M.A., Marks, R.J., I., Arabshahi, P., Gray, A.: Swarm intelligence for routing in communication networks. In: Global Telecommunications Conference, 2001. GLOBECOM '01. IEEE. vol. 6, pp. 3613–3617 (2001)
19. Kennedy, J., Eberhart, R.: Particle swarm optimization. In: IEEE International Conference on Neural Networks, pp. 1942–1948. Piscataway, NJ (1995)
20. Koziel, S., Michalewicz, Z.: Evolutionary algorithms, homomorphous mappings, and constrained parameter optimization. Evol. Comput. **7**(1), 19–44 (1999)
21. Michalewicz, Z., Attia, N.F.: Evolutionary optimization of constrained problems. In: Proceedings of the 3rd Annual Conference on Evolutionary Programming, pp. 98–108. World Scientific (1994)
22. Michalewicz, Z., Janikow, C.Z.: Handling constraints in genetic algorithms. In: Belew, R.K., Booker, L.B. (eds.) Proceedings of the Fourth International Conference on Genetic Algorithms (ICGA-91), pp. 151–157. University of California, San Diego, Morgan Kaufmann Publishers, San Mateo, California (1991)
23. Michalewicz, Z., Nazhiyath, G.: Genocop III: a co-evolutionary algorithm for numerical optimization with nonlinear constraints. In: Fogel, D.B. (ed.) Proceedings of the Second IEEE International Conference on Evolutionary Computation, pp. 647–651. IEEE Press, Piscataway, New Jersey (1995)

24. Michalewicz, Z., Schoenauer, M.: Evolutionary algorithms for constrained parameter optimization problems. Evol. Comput. **4**(1), 1–32 (1995)
25. Michalewicz, Z., Schoenauer, M.: Evolutionary algorithms for constrained parameter optimization problems. Evol. Comput. **4**(1), 1–32 (1996)
26. Omkar, S.N., Senthilnath, J., Khandelwal, R., Narayana Naik, G., Gopalakrishnan, S.: Artificial bee colony (ABC) for multi-objective design optimization of composite structures. Appl. Soft Comput. **11**(1), 489–499 (2011)
27. Pan, Q.K., Tasgetiren, M.F., Suganthan, P.N., Chua, T.J.: A discrete artificial bee colony algorithm for the lot-streaming flow shop scheduling problem. Information Sciences **181**(12), 2455–2468 (2011)
28. Powell, D., Skolnick, M.M.: Using genetic algorithms in engineering design optimization with non-linear constraints. In: Forrest, S. (ed.) Proceedings of the Fifth International Conference on Genetic Algorithms (ICGA-93). pp, 424–431. University of Illinois at Urbana-Champaign, Morgan Kaufmann Publishers, San Mateo, California (1993)
29. Richardson, J.T., Palmer, M.R., Liepins, G., Hilliard, M.: Some guidelines for genetic algorithms with penalty functions. In: Schaffer, J.D. (ed.) Proceedings of the Third International Conference on Genetic Algorithms (ICGA-89), pp. 191–197. George Mason University, Morgan Kaufmann Publishers, San Mateo, California (June 1989)
30. Schoenauer, M., Xanthakis, S.: Constrained GA optimization. In: Forrest, S. (ed.) Proceedings of the Fifth International Conference on Genetic Algorithms (ICGA-93), pp. 573–580. University of Illinois at Urbana-Champaign, Morgan Kauffman Publishers, San Mateo, California (1993)
31. Singh, A.: An artificial bee colony algorithm for the leaf-constrained minimum spanning tree problem. Appl. Soft Comput. (2008) (In Press)
32. Srinivas, N., Deb, K.: Muiltiobjective optimization using nondominated sorting in genetic algorithms. Evolut. Comput. **2**, 221–248 (1994)
33. Van Veldhuizen, D.A.: Multiobjective evolutionary algorithms: classifications, analyses, and new innovations. Ph.D. thesis, Wright Patterson AFB, OH, USA (1999), aAI9928483
34. Yang, C.K., Tsai, W.H.: Color image compression using quantization, thresholding, and edge detection techniques all based on the moment-preserving principle. Pattern Recogn. Lett. **19**, 205–215 (1998)
35. Zitzler, E., Thiele, L.: Multiobjective optimization using evolutionary algorithms—a comparative case study. In: Conference on Parallel Problem Solving from Nature (PPSN V), pp. 292–301. Amsterdam (1998)
36. Zitzler, E., Thiele, L., Laumanns, M., Fonseca, C., da Fonseca, V.: Performance assessment of multiobjective optimizers: an analysis and review. IEEE Trans. Evol. Comput. **7**(2), 117–132 (2003)
37. Zou, X., Chen, Y., Liu, M., Kang, L.: A new evolutionary algorithm for solving many-objective optimization problems. IEEE Trans. Syst. Man Cybern. Part B Cybern. **38**(5), 1402–1412 (2008)

Spider Monkey Optimization Algorithm

Harish Sharma, Garima Hazrati and Jagdish Chand Bansal

Abstract Foraging behavior of social creatures has always been a matter of study for the development of optimization algorithms. Spider Monkey Optimization (SMO) is a global optimization algorithm inspired by Fission-Fusion social (FFS) structure of spider monkeys during their foraging behavior. SMO exquisitely depicts two fundamental concepts of swarm intelligence: self-organization and division of labor. SMO has gained popularity in recent years as a swarm intelligence based algorithm and is being applied to many engineering optimization problems. This chapter presents the Spider Monkey Optimization algorithm in detail. A numerical example of SMO procedure has also been given for a better understanding of its working.

Keywords Spider monkey optimization · Swarm intelligence · Fission-fusion social structure · Numerical optimization

1 Spider Monkey Optimization

Spider monkey optimization (SMO) algorithm is a recent addition to the list of swarm intelligence based optimization algorithms [1, 2]. The update equations are based on Euclidean distances among potential solutions. The algorithm has extensively been applied to solve complex optimization problems. In [3], Dhar and Arora applied

H. Sharma · G. Hazrati
Rajasthan Technical University, Kota, India
e-mail: hsharma@rtu.ac.in

G. Hazrati
e-mail: ghazrati9@gmail.com

J. C. Bansal (✉)
South Asian University, New Delhi, India
e-mail: jcbansal@gmail.com

© Springer International Publishing AG, part of Springer Nature 2019
J. C. Bansal et al. (eds.), *Evolutionary and Swarm Intelligence Algorithms*, Studies in Computational Intelligence 779,
https://doi.org/10.1007/978-3-319-91341-4_4

Spider Monkey Optimization Algorithm (SMO) to design and optimize a fuzzy rule base. SMO is applied to solve optimal capacitor placement and sizing problem in IEEE-14, 30 and 33 test bus systems with the proper allocation of 3 and 5-capacitors by Sharma et al. [4]. In [5] Wu et al. introduced SMO for the synthesis of sparse linear arrays. The amplitudes of all the elements and the locations of elements in the extended sparse subarray are optimized by the SMO algorithm to reduce the side lobe levels of the whole array under a set of practical constraints. Cheruku et al. designed SM-RuleMiner for rule mining task on diabetes data [6]. The SMO has also been used to synthesize the array factor of a linear antenna array and to optimally design an E-shaped patch antenna for wireless applications [7].

The next part of the chapter details the motivation and working of the spider monkey optimization algorithm.

1.1 Motivation

Emergence of Fission-Fusion Society Structure

The concept of fission-fusion society is introduced by the biologist "Hans Kummer" while he was disentangling one of the most complex Mammalian Hamadryas baboons' social organization [8]. The competition for food among the group members of parent group when there is a shortage of food due to seasonal changes lead to fission into many groups and then fusion into a single group. When there is high availability of food then the group is largest whereas, in case of the smallest group, the food scarcity is at its peak. Fission part shows the food foraging behavior of spider monkeys and fusion represents combining of smaller groups to become a larger one.

Foraging Behavior of Spider Monkeys:

Spider monkeys live in the tropical rain forests of Central and South America and exist as far north as Mexico [9]. Spider monkeys are among the most intelligent New World monkeys. They are called spider monkeys because they look like spiders when they are suspended by their tails [10]. Spider monkeys always prefer to live in a unit group called 'parent group'. Based on the food scarcity or availability they split themselves or combine. Communication among them depends on their gestures, positions and whooping. Group composition is a dynamic property in this structure.

Social Organization and Behavior:

The social organization and behavior of spider monkeys can be understood through the following facts:

1. Spider monkeys live in a group of about 40–50 individuals.
2. All individuals in this community forage in small groups by going off in different directions during the day and everybody share the foraging experience in the night at their habitat.

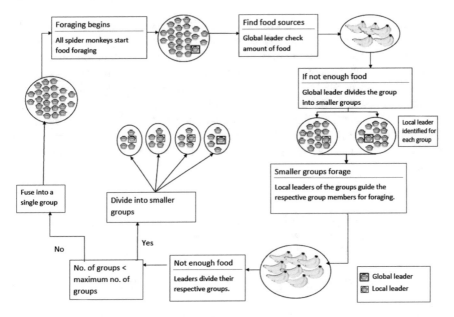

Fig. 1 Foraging behavior of Spider Monkeys

3. The lead female spider monkey decides the forage route.
4. If the leader does not find sufficient food then she divides the group into smaller groups and these groups forage, separately.
5. Individuals of the society might not be noticed closer at one place because of their mutual tolerance among each other. When they come into contact their gestures reflect that they are actually part of a large group.

Communication:

Spider monkeys share their intentions and observations using positions and postures. At long distances they interact with each other by particular sounds such as whooping or chattering. Each monkey has its own discernible sound by which other group members identify that monkey.

The above discussed foraging behavior of spider monkeys is shown in Fig. 1.

1.2   Spider Monkey Optimization Process

SMO is a meta-heuristic technique inspired by the intelligent foraging behavior of spider monkeys. The foraging behavior of spider monkeys is based on the fission-fusion social structure. Features of this algorithm depend on social organization of a group where a female leader takes decision whether to split or combine. The leader of the entire group is named here as the global leader while the leaders of the small groups are named as local leaders. With reference to the SMO algorithm, the phenomenon of food scarcity is defined by no improvement in the solution. Since SMO is a swarm intelligence based algorithm, each small group should have a minimum number of monkeys. Therefore, at any time if a further fission creating at least one group with less than the minimum number of monkeys, we define it as the time for fusion. In SMO algorithm, a Spider Monkey (SM) represents a potential solution. SMO consists of six phases: Local Leader phase, Global Leader phase, Local Leader Learning Phase, Global Leader Learning phase, Local Leader Decision phase and Global Leader Decision phase. All these phases of SMO are explained next:

Initialization:

In the initialization phase, SMO generates a uniformly distributed initial swarm of N spider monkeys, where SM_i represents the ith spider monkey (SM) in the swarm. Each SM_i is initialized as follows:

$$SM_{ij} = SM_{minj} + U(0, 1) \times \left(SM_{maxj} - SM_{minj}\right) \tag{1}$$

where, SM_{minj} and SM_{maxj} are lower and upper bounds of the search space in jth dimension and $U(0, 1)$ is a uniformly distributed random number in the range $(0, 1)$.

Local Leader Phase (LLP):

This is a vital phase of SMO algorithm. Here, all spider monkeys get chance to update themselves. Modification in the position of spider monkey is based on its local leader and local group members' experiences. The fitness value of each spider monkey is calculated at its new position and if fitness is higher than that of its old one, it gets updated otherwise not. Here, position update equation is:

$$SMnew_{ij} = SM_{ij} + U(0, 1) \times \left(LL_{kj} - SM_{ij}\right) + U(-1, 1) \times \left(SM_{rj} - SM_{ij}\right) \tag{2}$$

where, SM_{ij} is the jth dimension of ith SM, LL_{kj} represents the jth dimension of local leader of the kth group and SM_{rj} is the jth dimension of a randomly selected SM from the kth group such that $r \neq i$ and $U(-1, 1)$ is a uniformly distributed random number in the range $(-1, 1)$.

Here, it is clear from Eq. (2) that the spider monkey, which is going to update its position, is attracted towards the local leader while maintaining its self-confidence or persistence. The last component helps to introduce fluctuations in the search process, which helps to maintain the stochastic nature of the algorithm so that a premature stagnation can be avoided. A complete position update process of this phase is well explained in Algorithm 1. In this algorithm pr represents the perturbation rate for the current solution, whose value generally lies in range $[0.1, 0.8]$.

Algorithm 1: *Position update process in Local Leader Phase (LLP)*

---

for each member$SM_i \in k^{th}$ group **do**

    for each $j \in \{1, \dots, D\}$**do**

      if $U(0,1) \geq pr$**then**

$$SMnew_{ij} = SM_{ij} + U(0,1) \times \left(LL_{kj} - SM_{ij}\right) + U(-1,1) \times \left(SM_{rj} - SM_{ij}\right)$$

    else

    $SMnew_{ij} = SM_{ij}$

    end if

    end for

end for

Global Leader Phase (GLP):

After completing the local leader phase, the algorithm takes the step towards the global leader phase, Here solutions update is based on a selection probability, which is a function of the fitness. From objective function f_i the fitness fit_i can be calculated by Eq. (3).

$$fitness\ function = fit_i = \begin{cases} \frac{1}{1+f_i}, & if\ f_i \geq 0 \\ 1 + abs(f_i), & if\ f_i < 0 \end{cases} \tag{3}$$

The selection probability $prob_i$ is determined based on the roulette wheel selection. If fit_i is the fitness of ith SM then its probability of being selected in the global leader phase is calculated using either of the following two formulae:

$$prob_i = \frac{fitness_i}{\sum_{i=1}^{N} fitness_i}$$

or

$$prob_i = 0.9 \times \frac{fit_i}{max\_fit} + 0.1$$

To update the position, SM uses knowledge of the global leader, experience of neighboring SM and its own persistence. The position update equation in this phase is as follows:

$$SMnew_{ij} = SM_{ij} + U(0, 1) \times \left(GL_j - SM_{ij}\right) + U(-1, 1) \times \left(SM_{rj} - SM_{ij}\right) \tag{4}$$

where GL_j is the position of global leader in the jth dimension. This position update equation is divided into three components: the first component shows persistence of the parent (current) SM, the second component shows the attraction of the parent SM towards the global leader, and the last component is used to maintain the stochastic behavior of the algorithm. In this equation, the second component is used to enhance the exploitation of already identified search space, while the third component helps the search process to avoid premature convergence or to reduce the chance of being stuck in a local optima. The whole search process of this phase is described in Algorithm 2 below:

Algorithm 2: *Position update process in Global Leader Phase (GLP)*

$count = 0$;
while $count < group$ size **do**
for each member $SM_i \in$ group **do**
    if $U(0,1) < prob_i$ **then**
    $count = count + 1$
    Randomly select $j \in \{1, \dots, D\}$
    Randomly select $SM_r \in$ group s.t. $r \neq i$
    $SMnew_{ij} = SM_{ij} + U(0,1) \times \left(GL_j - SM_{ij}\right) + U(-1,1) \times \left(SM_{rj} - SM_{ij}\right)$
    end if
  end for
end while

It is clear from Algorithm 2 that the chance of updating a solution depends on $prob_i$. Therefore, the solution of high fitness will get more chance as compared to less fit solutions to update its position. Further, a greedy selection approach is applied to the updated solutions i.e., out of the updated and previous SM, the better fit solution is considered.

Global Leader Learning Phase:

In this phase, the algorithm finds the best solution of the entire swarm. The identified SM is considered as the global leader of the swarm. Further, the position of the global leader is checked and if it is not updated then the counter associated with the global leader, named as Global Limit Count (GLC), is incremented by 1, otherwise it is set to 0.

Global Limit Count is checked for global leader and is compared with Global Leader Limit (GLL).

Local Leader Learning Phase:

In this segment of the algorithm, the position of the local leader gets updated by applying a greedy selection among the group members. If local leader doesn't update its position then a counter associated with local leader, called Local Limit Count (LLC) is incremented by 1; otherwise the counter is set to 0. This process is applied to every group to find its respective local leader.

Local Limit Count is a counter that gets incremented till it reaches a fix threshold called Local Leader Limit (LLL).

Local Leader Decision Phase:

Before this phase, local leaders and the global leader have been identified. If any local leader does not get reorganized to a particular verge, known as Local Leader Limit, then all the members of that group update their positions either by random initialization or by using global leader's experience via Eq. (5). Equation (5) is applied with a probability pr called the perturbation rate.

$$SMnew_{ij} = SM_{ij} + U(0, 1) \times (GL_j - SM_{ij}) + U(0, 1) \times (SM_{rj} - LL_{kj}) \quad (5)$$

From this equation, it can be understood that the solutions of this group are repelled from the existing local leader as it is exhausted (not updated up to LLL number of iterations) and solutions are attracted towards the global leader to change the existing search directions and positions. Further, based on pr, some dimensions of the solutions are randomly initialized to introduce some disturbance in the existing positions of the solutions. Here, Local Leader Limit is the parameter which checks that local leader does not get stuck in local minima, and normally, it is calculated as D × N, where D is dimension and N is total number of SM. If LLC is more than LLL then LLC is set to zero and SM gets initialized as described above to improve the exploration of the search space.

The process of local leader decision phase is described in Algorithm 3.

Algorithm 3: *Local Leader Decision Phase (LLD):*

If $LocalLimitCount > LocalLeaderLimit$ **then**
    $LocalLimitCount = 0$
for each $j \in \{1, \dots, D\}$ **do**
    if $U(0,1) \geq pr$ **then**
        $SMnew_{ij} = SM_{minj} + U(0,1) \times (SM_{maxj} - SM_{minj})$
    else
        $SMnew_{ij} = SM_{ij} + U(0,1) \times (GL_j - SM_{ij}) +$
                                   $U(0,1) \times (SM_{rj} - LL_{kj})$
    end if
end for
end if

Global Leader Decision Phase:

Similar to the local leader decision phase, if the global leader does not get reorganized to a particular verge known as Global leader limit, then the global leader divides the swarm into smaller groups or fuse groups into one unit group. Here, GLL is the parameter, which check whether there is any premature convergence, and varies in the range of N/2 to 2 × N. If GLC is more than GLL then GLC is set to zero and number of groups are compared to maximum groups. If existing number of groups is less than the pre-defined maximum number of groups then the global leader further divides the groups otherwise combines to form a single or parent group. This fission-fusion process is described in Algorithm 4.

Algorithm 4 *Global Leader Decision Phase (GLD):*

if $GlobalLimitCount > GlobalLeaderLimit$ **then**
    $GlobalLimitCount = 0$
if $Number\ of\ groups < MG$ **then**
    Divide the swarms into groups
else
    Combine all the groups to make a single group
end if
    update Local Leaders position
end if

Following algorithm 5 provides the complete working mechanism of SMO to solve an optimization problem.

Algorithm 5: *Spider Monkey Optimization*

Step 1.	**Initialize** population, local leader limit, global leader limit and perturbation rate pr;
Step 2.	**Evaluate** the population;
Step 3.	**Identify** global and local leaders;
Step 4.	**Position update** by local leader phase (Algorithm 1);
Step 5.	**Position update** by global leader phase (Algorithm 2);
Step 6.	**Learning** through global leader learning phase;
Step 7.	**Learning** through local leader learning phase;
Step 8.	**Position update** by local leader decision phase (Algorithm 3);
Step 9.	**Decide fission or fusion** using global leader decision phase (Algorithm 4);
Step 10.	If termination condition is satisfied stop and declare the global leader position as the optimal solution else go to step 4.

---

2 Analyzing SMO

SMO better balances between exploitation and exploration while search for the optima. Local Leader phase is used to explore the search region as in this phase all the members of the groups update their positions with high perturbation in the dimensions. While the global leader phase promotes the exploitation as in this phase, better candidates get more chance for updating their positions. This property makes SMO a better candidate among the search based optimization algorithms. SMO also possesses an inbuilt mechanism for stagnation check. Local leader learning phase and global leader learning phase, are used to check if the search process is stagnated. In case of stagnation (at local or global level) local leader and global leader decision phases work. The local leader decision phase creates an additional exploration while in the global leader decision phase, a decision about fission or fusion is taken. Therefore, in SMO exploration and exploitation are better balanced while maintaining the convergence speed.

3   Parameters of SMO

SMO has mainly four new control parameters: Local leader limit, Global leader limit, the maximum number of groups (MG), and perturbation rate pr. The suggested parameter setting are given as follows [2]:

- $MG = \frac{N}{10}$, i.e., it is chosen in such a way such that the minimum number of SM's in a group should be 10,
- Global leader limit $\in \left[\frac{N}{2}, 2 \times N\right]$,
- Local leader limit is set to $D \times N$,
- $pr \in [0.1, 0.8]$.

4   Performance Analysis of SMO

Performance of SMO has been analyzed against three well-known meta-heuristics, Artificial Bee Colony (ABC), Differential Evolution (DE), and Particle Swarm Optimization (PSO) in [2]. After testing on 25 benchmark problems and performing various statistical tests, it is concluded that the SMO is a competitive meta-heuristic for optimization. It has been shown that the SMO performed well for unimodal, multimodal, separable and non-separable optimization problems [2]. It was found that for continuous optimization problems SMO should be preferred over PSO, ABC or DE for better reliability.

5   A Worked-Out Example

This section describes a numerical example of SMO. In this step-by-step procedure of SMO, a simple optimization problem $f(x) = x_1^2 + x_2^2$ is solved using the SMO algorithm.

Consider an optimization problem:

$$\text{Minimize } (x) = x_1^2 + x_2^2; \quad -5 \le x_1, x_2 \le 5$$

Control parameters of SMO:

Swarm or population size, N = 20.
Dimension of problem, D = 2.
If we consider minimum number of individuals in a group are 10 then the maximum number of groups (MG) = $\frac{N}{10} = 2$.
Global Leader Limit, $GLL \in \left\{\frac{N}{2}, 2N\right\} \Rightarrow GLL \in \{10, 40\}$.
Let GLL = 30.
Local Leader Limit, $LLL = D \times N = 2 \times 20 = 40$.

Perturbation rate, $pr \in [0.1, 0.8]$.
Let $pr = 0.7$.

Initialization:

Now, we randomly initialize positions (SMs) of 20 food resources in the range of $[-5, 5]$.

SM number	x_1	x_2	SM number	x_1	x_2
1	1.4	1.2	11	0.1	−0.9
2	−2.4	−2.5	12	0.3	0.3
3	0.6	−0.4	13	−0.4	0.6
4	0.3	1.5	14	0.5	0.7
5	−0.7	1.9	15	1.3	−1.5
6	2.9	3.2	16	−1.1	0.8
7	1.6	−0.9	17	0.8	−0.9
8	0.8	0.2	18	0.4	−0.2
9	−0.5	0.1	19	−0.6	0.3
10	0.3	0.2	20	0.8	1.6

Corresponding function values are

SM number	$f_i(x)$	SM number	$f_i(x)$
1	3.4	11	0.82
2	12.01	12	0.18
3	0.52	13	0.52
4	2.34	14	0.74
5	4.1	15	3.94
6	18.65	16	1.85
7	3.37	17	1.45
8	0.68	18	0.2
9	0.26	19	0.45
10	0.13	20	3.2

Fitness values are calculated by using the formula of fitness function, i.e.

$$fitness\ function = fit_i = \begin{cases} \frac{1}{1+f_i(x)}, & if\ f_i(x) \geq 0 \\ 1 + abs(f_i(x)), & if\ f_i(x) < 0 \end{cases}$$

Here since the maximum fitness value is 0.8850, which corresponds to 10th SM, 10th SM becomes the global leader. At this stage there is only single group so 10th SM is the global as well as local leader.

SM number	fit_i	SM number	fit_i
1	0.227	11	0.549
2	0.077	12	0.847
3	0.658	13	0.658
4	0.299	14	0.575
5	0.196	15	0.202
6	0.051	16	0.351
7	0.229	17	0.408
8	0.595	18	0.833
9	0.794	19	0.690
10	0.885	20	0.238

Position Updated Phases:
Local Leader Phase

In this phase, all the SMs will get a chance to update their positions. For updating the position, following position update equation is used.

$$SMnew_{ij} = SM_{ij} + U(0, 1) \times \left(LL_{kj} - SM_{ij}\right) + U(-1, 1) \times \left(SM_{rj} - SM_{ij}\right)$$

where,

$SMnew_{ij}$ = new position of ith spider monkey in the jth direction.
SM_{ij} = old position of ith spider monkey in the jth direction.
SM_{rj} = position of randomly selected spider monkey in the jth direction.
LL_{kj} = position of the local leader. Since there is only one group, $k = 1$.

Updating 1st spider monkey (i=1)

For first dimension $j = 1$, generate a random number U(0,1). Let U(0, 1) = 0.3. Since pr (=0.7) $\leq$ U(0, 1) is false, therefore $SMnew_{11} = SM_{11}$.
For $j = 2$, let U(0, 1) = 0.8. Since $pr \leq 0.8$, SM_{12} is going to update.
If randomly selected neighboring solution index r is 6 and U(-1, 1) = -0.7 then

$$SMnew_{12} = 1.2 + 0.8 \times (0.2 - 1.2) + (-0.7) \times (3.2 - 1.2)$$
$$SMnew_{12} = -1$$

So the new solution $x_1 = (1.4, -1)$.
Calculating function value and fitness of $SMnew_1$
$f_1(SMnew_1) = 2.96$; fit $(SMnew_1) = 0.252$.
Applying greedy selection between $SMnew_1$ and SM_1 based on fitness
$0.227 < 0.252$ so it is improved, i.e. $SM_1 = (1.4, -1)$.
Similarly other solutions are also updated and listed next.
For Global Leader Phase, we need to calculate probability function using fitness vector, i.e.

SM number	Updated dimension j	SM_{new}		$f_i(x)$	Fit_i
1	2	1.4	−1	2.96	0.252
2	1	−1.56	−2.5	8.6836	0.1032
3	2	0.6	0.12	0.3744	0.727
4	–	0.3	1.5	2.34	0.299
5	1	−0.34	1.5	2.366	0.2971
6	–	0.69	3.2	10.716	0.0854
7	1	1.6	−0.9	3.370	0.2288
8	–	0.4	0.2	0.200	0.8333
9	–	−0.5	0.1	0.260	0.7937
10	–	0.3	−0.2	0.130	0.8850
11	2	0.1	0.31	0.106	0.9041
12	1	0.42	0.3	0.266	0.7896
13	–	−0.4	0.6	0.520	0.6579
14	2	0.5	−0.26	0.318	0.7590
15	–	1.3	−1.5	3.940	0.2024
16	2	−1.1	−0.165	1.237	0.4470
17	2	0.8	−0.33	0.749	0.5718
18	2	0.4	0.14	0.180	0.8477
19	–	−0.6	0.3	0.450	0.6897
20	2	0.8	−0.142	0.660	0.6024

$$prob_i = 0.9 \times \frac{fit_i}{max\_fit} + 0.1 \; ; \; \text{Here}, max\_fit = max\{fit_i; \; i = 1, 2, \ldots, 20\}$$

Clearly, $max\_fit = 0.9041$, which corresponds to the 11th solution. Following Table lists the fitness probabilities.

SM	$prob_i$	SM	$prob_i$
1	0.38968	11	1
2	0.202732	12	0.886019
3	0.823703	13	0.754916
4	0.397644	14	0.855558
5	0.395753	15	0.301482
6	0.185013	16	0.544973
7	0.327762	17	0.669207
8	0.929521	18	0.943856
9	0.890101	19	0.786572
10	0.980987	20	0.699668

Global Leader Phase

In this phase, SMs are updated based on the probability $prob_i$. As $prob_i$ is the function of fitness, high fit SMs will get more chance for the update.

In this phase, the position update equation is

$$SMnew_{ij} = SM_{ij} + U(0, 1) \times (GL_j - SM_{ij}) + U(-1, 1) \times (SM_{rj} - SM_{ij})$$

where, GL_j =Global leader position in jth direction.

The total number of updates in Global Leader Phase depends on population size in the group. In this example, we are showing position update process of two spider monkeys (8th and 17th) only. It should be noted that in this phase also, only one dimension of every selected solution is updated.

Updating 8th spider monkey ($i = 8$)

$prob_8 = 0.929521$. Let $U(0, 1) = 0.6 < prob_8$. Therefore 8th SM is to be updated.

Let $j = 2$ selected randomly.

Apply the position update equation of global leader phase to $SM_8 = (0.4, 0.2)$.

We obtain $SMnew_8 = (0.4, -0.75)$.

$f_8(x) = 0.7225$ and $fit_8 = 0.5805$.

Apply the greedy selection between $SMnew_8$ and SM_8 based on fitness.

Since $0.8333 > 0.5805$, SM_8 is not updated.

Updating 17th spider monkey ($i = 17$)

$prob_{17} = 0.6692$. Let $U(0, 1) = 0.52 < prob_{17}$. Therefore 17th SM is to be updated.

Let $j = 1$ is selected randomly.

Apply the position update equation of global leader phase to $SM_{17} = (0.8, -0.33)$.

We obtain $SMnew_{17} = (-0.264, -0.33)$.

$f_{17}(x) = 0.1785$ and $fit_{17} = 0.8484$.

Apply the greedy selection between $SMnew_{17}$ and SM_{17} based on fitness.

Since $0.571 < 0.8484$, SM_{17} is improved.

Updated $SM_{17} = (-0.264, -0.33)$.

After first round (all the solutions should get chance to update their positions), the new positions of the SMs are as follows:

As the total 12 solutions have been updated in first round of global leader phase and as per the termination condition of this phase, the total number of modifications should be equal to the number of SMs in the population. Hence, a next round will be initiated to update the SMs. After the second round, the updated positions of the SMs are as follows:

Now, in this phase, the total numbers of SM modifications are 20, so the phase is terminated now. Hence, it is clear from this phase that the solutions having high fitness value get more chance to update their positions, which improves the exploitation capability of the algorithm.

SM number	j	SM_{newij}		$f_i(x)$	Fit_i
1	–	1.4	−1	2.96	0.252525
2	–	−1.56	−2.5	8.6836	0.103267
3	2	0.6	0.4	0.52	0.657895
4	–	0.3	1.5	2.34	0.299401
5	1	0.1	1.5	2.26	0.306748
6	–	0.69	3.2	10.7161	0.085353
7	2	1.6	−0.3	2.65	0.273973
8	1	1.1	0.2	1.25	0.444444
9	1	−0.8	0.1	0.65	0.606061
10	2	0.3	−0.9	0.9	0.526316
11	2	0.1	−0.4	0.17	0.854701
12	1	0.3	0.3	0.18	0.847458
13	1	−0.8	0.6	1	0.5
14	1	−0.2	−0.26	0.1076	0.902853
15	–	1.3	−1.5	3.94	0.202429
16	–	−1.1	−0.165	1.237225	0.446982
17	1	−0.264	−0.33	0.178596	0.848467
18	1	0.45	0.14	0.2221	0.818264
19	–	−0.6	0.3	0.45	0.689655
20	–	0.8	−0.142	0.660164	0.60235

Global Leader Learning Phase

Global leader learning phase decides the global leader of the swarm. All solutions' fitness will be compared with each other. If the global leader attains the better position than global limit count is set to 0 otherwise it is incremented by 1. Since the 9th SM's fitness is the best in the updated swarm, it becomes the global leader. Also the global limit count will be set to 0 as the global leader has been updated.

Local Leader Learning Phase

Local leader learning phase decides the local leaders of the groups. Similar to global leader learning phase all solutions' fitness will be compared with each other. If the local leader attains the better position than local limit count is set to 0 otherwise it is incremented by 1. Here we have only one group so 9th SM is the global leader as well as the local leader. The local limit count is set to 0 as the local leader has been updated.

Local Leader Decision Phase

As per our parameter settings local leader limit is 40. Since local leader limit count $= 0 <$ local leader limit $= 40$, this phase will not be implemented.

Global Leader Decision Phase

SM number	j	SM_{newij}		$f_i(x)$	Fit_i
1	–	1.4	−1	2.96	0.252525
2	–	−1.56	−2.5	8.6836	0.103267
3	–	0.6	0.4	0.52	0.657895
4	–	0.3	1.5	2.34	0.299401
5	–	0.1	1.5	2.26	0.306748
6	–	0.69	3.2	10.7161	0.085353
7	–	1.6	−0.3	2.65	0.273973
8	–	1.1	0.2	1.25	0.444444
9	1	−0.4	0.1	0.17	0.854701
10	1	−0.8	−0.9	1.45	0.408163
11	1	1.2	−0.4	1.6	0.384615
12	1	1.8	0.3	3.33	0.230947
13	–	−0.8	0.6	1	0.5
14	2	−0.2	0.7	0.53	0.653595
15	–	1.3	−1.5	3.94	0.202429
16	–	−1.1	−0.165	1.237225	0.446982
17	2	−0.264	−0.8	0.709696	0.584899
18	2	0.45	−0.3	0.2925	0.773694
19	1	−0.4	0.3	0.25	0.8
20	–	0.8	−0.142	0.660164	0.60235

In this phase, the position of the global leader is monitored and if it is not updated up to the global leader limit (=30, for this example), then the population is divided into smaller groups. If the numbers of sub-groups are reached to its maximum count (=2, for this example) then all sub-groups are combined to form a single group. After taking the decision, the global limit count is set to 0 and the positions of the local leaders are updated.

In order to explain the role of global leader decision phase, consider a case when for some iteration, global limit count is 31, then the group is divided into two sub-groups. The solutions SM_1–SM_{10} belong to the first group while SM_{11}–SM_{20} belong to the second group. Let the swarm is represented by the following Table:

As the fitness of the 9th SM is highest among all the SMs of the first group, so it is designated as the local leader of the first group, i.e. $LL_1 = (-0.4, 0.1)$. Further, for the second group 19th SM has the highest fitness, so it is considered the local leader of the second group, i.e. $LL_2 = (-0.4, 0.3)$. The local limit counts for both the local leaders are set to 0.

Also it can be seen that the fitness of the 9th SM is the best among all the members of the swarm, hence the 9th SM is considered as the global leader of the swarm, i.e. $GL = (0.1, 0.31)$. As the global leader decision has been implemented, the global limit count becomes 0.

Group number, SM number	SM_{newij}		$f_i(x)$	Fit_i
K=1, SM=1	1.4	−1	2.96	0.252525
K=1, SM=2	−1.56	−2.5	8.6836	0.103267
K=1, SM=3	0.6	0.4	0.52	0.657895
K=1, SM=4	0.3	1.5	2.34	0.299401
K=1, SM=5	0.1	1.5	2.26	0.306748
K=1, SM=6	0.69	3.2	10.7161	0.085353
K=1, SM=7	1.6	−0.3	2.65	0.273973
K=1, SM=8	1.1	0.2	1.25	0.444444
K=1, SM=9	−0.4	0.1	0.17	0.854701
K=1, SM=10	−0.8	−0.9	1.45	0.408163
K=2, SM=11	1.2	−0.4	1.6	0.384615
K=2, SM=12	1.8	0.3	3.33	0.230947
K=2, SM=13	−0.8	0.6	1	0.5
K=2, SM=14	−0.2	0.7	0.53	0.653595
K=2, SM=15	1.3	−1.5	3.94	0.202429
K=2, SM=16	−1.1	−0.165	1.237225	0.446982
K=2, SM=17	−0.264	−0.8	0.709696	0.584899
K=2, SM=18	0.45	−0.3	0.2925	0.773694
K=2, SM=19	−0.4	0.3	0.25	0.8
K=2, SM=20	0.8	−0.142	0.660164	0.60235

After global leader decision phase, the swarm is updated by local leader phase and other phases in the similar way. This process is continued iteratively, until the termination criteria is reached.

6 Conclusion

In this chapter, a recent swam intelligence based algorithm, namely Spider Monkey Optimization is discussed which is developed by taking inspiration from the social behavior of spider monkeys. In SMO, the local leader phase and the global leader phase help in exploitation of the search space, while exploration is done through the local leader decision phase and global leader decision phase. SMO performance analysis shows that SMO outpaced ABC, DE and PSO, in terms of dependability, effectiveness and precision. However, the presence of a large number of user dependent parameters in SMO is a matter of concern for further research. Self-adaptive parameter tuning may help to improve the robustness and reliability of the algorithm. Just in 4 years, a large number of publications on development and applications of SMO show that it has a great potential to be an efficient optimizer.

Note: SMO codes in C++, Python and Matlab can be downloaded from http://smo.scrs.in/.

References

1. Bonabeau, E., Dorigo, M., Theraulaz, G.: Swarm Intelligence: From Natural to Artificial Systems. Oxford University Press, New York, NY (1999)
2. Bansal, J.C., Sharma, H., Jadon, S.S., Clerc, M.: Spider monkey optimization algorithm for numerical optimization. Memetic Comput. **6**(1), 31–47 (2014)
3. Dhar, J., Arora, S.: Designing fuzzy rule base using spider monkey optimization algorithm in cooperative framework. Future Comput. Info. J. **2**(1), 31–38 (2017). ISSN 2314-7288
4. Sharma, A., Sharma, H., Bhargava, A., et al.: Memetic Comp. **9**, 311 (2017). https://doi.org/10.1007/s12293-016-0208-z
5. Wu, H., Yan, Y., Liu, C., Zhang, J.: Pattern synthesis of sparse linear arrays using spider monkey optimization. In: IEICE Transactions on Communications, Released 01 March 2017
6. Cheruku, R., Edla, D.R., Kuppili, V.: SM-RuleMiner: Spider monkey based rule miner using novel fitness function for diabetes classification. Comput. Biol. Med. **81**, 79–92 (2017)
7. Al-Azza, A.A., Al-Jodah, A.A., Harackiewicz, F.J.: Spider monkey optimization: a novel technique for antenna optimization. IEEE Antennas Wirel. Propag. Lett. **15**, 1016–1019 (2016)
8. Couzin, I.D, Laidre, M.E: Fission–fusion populations. Curr. Biol. **19**(15), R633–R635
9. https://www.nationalgeographic.com/animals/mammals/group/spider-monkeys/
10. https://animalcorner.co.uk/animals/spider-monkey/

Genetic Algorithm and Its Advances in Embracing Memetics

Liang Feng, Yew-Soon Ong and Abhishek Gupta

Abstract A Genetic Algorithm (GA) is a stochastic search method that has been applied successfully for solving a variety of engineering optimization problems which are otherwise difficult to solve using classical, deterministic techniques. GAs are easier to implement as compared to many classical methods, and have thus attracted extensive attention over the last few decades. However, the inherent randomness of these algorithms often hinders convergence to the exact global optimum. In order to enhance their search capability, learning via memetics can be incorporated as an extra step in the genetic search procedure. This idea has been investigated in the literature, showing significant performance improvement. In this chapter, two research works that incorporate memes in distinctly different representations, are presented. In particular, the first work considers meme as a local search process, or an individual learning procedure, the intensity of which is governed by a theoretically derived upper bound. The second work treats meme as a building-block of structured knowledge, one that can be learned and transferred across problem instances for efficient and effective search. In order to showcase the enhancements achieved by incorporating learning via memetics into genetic search, case studies on solving the NP-hard capacitated arc routing problem are presented. Moreover, the application of the second meme representation concept to the emerging field of evolutionary bilevel optimization is briefly discussed.

L. Feng (✉) · Y.-S. Ong · A. Gupta
School of Computer Engineering, Nanyang Technological University,
Singapore, Singapore
e-mail: feng0039@ntu.edu.sg

Y.-S. Ong
e-mail: asysong@ntu.edu.sg

A. Gupta
e-mail: abhishekg@ntu.edu.sg

L. Feng
College of Computer Science, Chongqing University,
Chongqing, China
e-mail: liangf@cqu.edu.cn

© Springer International Publishing AG, part of Springer Nature 2019
J. C. Bansal et al. (eds.), *Evolutionary and Swarm Intelligence
Algorithms*, Studies in Computational Intelligence 779,
https://doi.org/10.1007/978-3-319-91341-4_5

Keywords Evolutionary optimization · Genetic algorithm · Memetics

1 Introduction

Evolutionary Algorithms (EAs) serve as appealing optimisation strategies for solving non-linear programming problems characterized by non-convex, disjoint and noisy objective functions. Unlike conventional numerical optimisation methods, evolutionary algorithms produce new design points that do not use information about the local slope of the objective function and are thus not prone to stalling at local optima. The Genetic algorithm (GA) [1, 2], as one of the most popular EAs, is inspired by Darwin's survival of the fittest strategy, with sexual reproduction and Mendel's theory of genetics as the basis of biological inheritance. GA maintains a population of solutions while making use of competitive selection, recombination and mutation operators to generate new solutions which are biased towards better regions of the search space. Their popularity also lies in the ease of implementation and their ability to converge close to the global optimum. However, a naive GA will usually exhibit very slow convergence properties and/or rapid diversity loss (i.e. convergence to a local optimum only). These problems often limit the practicality of GAs on many complex real world problems where a good solution is needed within limited computational time budget.

Today, it is well recognized that the processes of learning and the transfer of what has been learned are pivotal tools for humans in effective and efficient problem-solving [3]. Learning is the fundamental necessity for humans to function and adapt to the fast evolving society. In the literature, memetics has been established as a paradigm to incorporate learning into computational intelligence for problem-solving [4, 5]. In memetics, the term "meme" has been defined as "the basic unit of cultural transmission via imitation" [6]. Like genes that serve as "instructions for building proteins", memes are then "instructions for carrying out behavior, stored in brains". In recent years, the science of memetics has spread across the fields of biology, cognition, psychology, etc., and attracted significant attention.

In the context of evolutionary optimization, particularly the GA, memes have been typically perceived as either being individual learning procedures [7–9] or as being useful traits for efficient problem-solving [10–12]. When meme is perceived as an individual learning procedure, adaptive improvement procedure or local search operator, meme is used to enhance the convergence capability of GAs by locally refining a found solution and updating the population accordingly. This integration has been established as an extension of the canonical GA, with nomenclatures such as hybrid, adaptive hybrid or Memetic Algorithm (MA), assigned in the literature. On the other hand, some genetic search algorithms incorporate memes by treating them as useful traits, structured knowledge or building blocks of problems. These approaches are meme-centric computing paradigms for search, where meme is the fundamental building block of cultural evolution that can be learned from problem-solving experiences and transmitted across problems to bias the genetic search for efficient and effective problem-solving on unseen problems.

In this chapter, our focus is on GAs and their advancements to incorporate learning through memetics as powerful problem-solving strategies. To this end, two specific memetic concepts have been presented herein. The first one is a formal probabilistic memetic framework, which embeds individual learning into genetic algorithm, and governs, at runtime, whether genetic search or individual learning should be favored, with a theoretical upper bound on the intensity of the individual learning process. It is worth noting that existing strategies for controlling the MA parameters are mostly designed based on heuristics measures and come with little theoretical motivation. Subsequently, a meme-centric framework: genetic algorithm + transfer learning for search, one that models how a human solves problems, is presented. The recent success of transfer learning, which uses data from a related source task to augment learning in a new or target task, has been incorporated into genetic search via memetics in this work. Particularly, the structured knowledge or useful traits in the form of memes have been identified, learned and transmitted across problem instances for enhanced genetic searches.

The remainder of this chapter is organized as follows. In Sect. 2, we begin with a brief overview of GAs and their advances in incorporating memes as either local search processes or structured knowledge. Section 3 describes the formal probabilistic memetic framework, which balances the genetic search and local search process by a theoretical upper bound. Subsequently, the meme-centric framework: genetic algorithm + transfer learning and its specific realization for solving routing problem is presented in Sect. 4. The case studies of both, the probabilistic memetic framework and meme-centric framework, on capacitated arc routing problem data sets are presented in Sect. 5. Section 6 presents a brief discussion on the application of meme as a transfer learning entity to bilevel programming problems which are prevalent in a wide variety of real world situations. Finally, in Sect. 7 a summary of the central ideas of the chapter is provided and the main conclusions are drawn.

2 Preliminary

In this section we first present a brief introduction to GAs. Subsequently, an overview of the advances in GAs by incorporating memes in the search are presented.

2.1 Genetic Algorithm

The development of Genetic Algorithm (GA) is inspired by the survival of the fittest principle in the Darwinian theory of natural evolution, and Mendel's theory of genetics. The GA search begins with a set of solutions, often labeled as a population. Each solution is represented as a chromosome in the population. In each generation, the reproduction operators such as mutation and crossover are used for the creation of the new chromosomes. The performance or suitability of a chromosome is then defined by some fitness. This fitness value of a chromosome serves as the basis for its survival into the next generation, i.e., fitter chromosome has higher chances of surviving. This fitness based selection mechanism ensures the fitter chromosomes to survive across generations, whereas the least fit ones fails to replicate well. The process of evolution is repeated until some stopping condition is satisfied.

The first design issue to consider when developing a GA is the encoding representation of the chromosome. Binary encodings as introduced by Holland [13, 14] is among the first and most commonly used in genetic algorithms. To date, there have been many new extensions to the basic binary encoding schema, and these include gray coding, real coding and the Hillis's diploid binary encoding scheme, etc. Other specialized encoding schemes have also been proposed for specific disciplines to enhance search efficiency.

Upon deciding on an encoding representation, the GA operators will be conducted. Algorithm 1 outlines the basic steps in a Genetic Algorithms, followed by detailed description on each GA operator.

Algorithm 1 Genetic Algorithm

Initialization: Generate an initial population
While Stopping conditions are not satisfied **do**
      Evaluate all individuals in the population
      Select all individuals in the population
      Apply genetic operators (crossover, mutation) to generate offsprings
      Replace a proportion or the entire population with the offsprings
End While

Initialization: To start a genetic algorithm search, the first step is to initialize a population of solutions or individuals of the problem of interest. Traditionally, the population is generated randomly, allowing the entire range of possible solutions (the search space). The population size depends on the nature of the problem, but typically contains dozens or several hundreds or thousands of possible solutions. However, to narrow down the search space and make the search more focus, approaches use heuristic search methods to initialize the population of GA are also proposed in the literature.

Selection: The selection operator denotes the reproduction operator that define the chromosomes that survives to form the population in the subsequent generations. Inspired by Darwin's evolutionary theory of natural selection, the fitter chromosomes shall survive better through the generations. Let $P(x)$ be the probability that an individual x in the current population is selected to reproduce the next population, there are several popular selection schemes [13, 14]:

– Roulette Wheel Selection: It is also known as fitness proportionate selection. In roulette wheel selection, each individual is selected according to its fitness. The fitter the chromosomes are, the higher are the chances of being selected and survive to the next generation. In particular, the probability $P(x)$ of selecting a chromosome or individual x in the next generation can be computed as follows:

$$P(x) = \frac{f(x)}{\sum_{x=1}^{n} f(x)} \tag{1}$$

where $f(x)$ denote the fitness value of individual x.

- Rank Selection: In rank selection, the individuals are sorted by their fitness. The probability for individual x to be selected is then inversely proportional to its position in this sorted list. In another word, individual at the head of the list is more likely to be selected than the next individual down the list. This selection approach addresses the problem arisen in roulette wheel selection, that when fitness differs very much among individuals, the individual with higher fitness will always be selected.
- Stochastic Universal Selection: It is an extension of the roulette wheel selection to avoid the limitation of selecting the same individuals repeatedly when elite individuals dominant, i.e., fitness values that significantly superior to the others in the population. In particular, it uses a single random value to sample all of the solutions by choosing them at evenly spaced intervals. In such a way, weaker individuals with lower fitness value have more chance to be chosen.
- Tournament Selection: It first selects a number of individuals (two or more) from the population at random. A tournament among the individuals is then carried out and only the best of those individuals survive to the next generation, the other(s) is then returned to the population for the subsequent selection. The selective pressure of tournament selection can be adjusted by means of the tournament size parameter, which makes it a flexible selection scheme. Note that a tournament size of 1 would be equivalent to selecting individuals at random, and a tournament size equal to the population size is equivalent to selecting the best individual at any given point.

Crossover: Crossover is a reproduction operator that combines (mates) two or more chromosomes (parents) to produce a new chromosome (offspring). The idea behind crossover is that the new chromosome may serve better than both of the parents via inheriting their superior characteristics. Usually, crossover operator randomly chooses a point and exchanges the sub-sequences before and after that point between two individuals to create two offsprings. Clearly, the choice of crossover operator depends on the encoding scheme. Generally, there are three types of crossover operator which are applicable for both binary and real codding chromosomes, namely single-point crossover, two-point crossover and uniform crossover. An illustration of single-point crossover is depicted in Fig. 1. Other types of crossover may include the cut and splice crossover, half uniform crossover arithmetic crossover, heuristic crossover, etc., that can refer to [1, 14].

Among the various crossover operators, single-point crossover is the simplest form. A single crossover position is chosen at random and portions of two parents after the crossover position are exchanged to create two new offsprings. In two-point crossover, two positions are randomly selected and the respective segments of the

Fig. 1 One-point crossover operation of the genetic algorithm ('—' is the crossover point)

Chromosome A = 10001|10101110000
Chromosome B = **10001**|**01110011011**
Child A = 10001|**01110011011**
Child B = **10001**|10101110000

Fig. 2 Flip-bit mutation operation of the genetic algorithm (Bits selected for mutation are shown in bold with underline)

Chromosome A = 1000110101110000
Chromosome B = 1000101110011011
Mutated Chromosome A = 1000111101110000
Mutated Chromosome B = 1000101110010011

parents are swapped. Generally, two point crossover is better than one-point because it minimizes schema disruption better. In uniform crossover, parents bits are chosen at random for swapping. Uniform crossover is generally more disruptive than one-point crossover. However, its disruptive character might be helpful in overcoming premature convergence to local optima.

Mutation: Mutation is a reproduction operator that alters one ore more gene values in the chromosome. It randomly modifies the genes of an individual, subjected to a small mutation probability. A core mechanism of mutation is to induce some level of diversity into the population to reduce the effect of getting trapped in local optimum during the GA search. Similar to the crossover operator, mutation operators are encoding scheme dependent. With binary encoding for example, the mechanism of the mutation operation can be depicted using Fig. 2. Other forms of established mutation operators would include the flip bit mutation, non-uniform mutation, uniform mutation, gaussian mutation, etc [14–16].

Replacement: Replacement denotes the process that replace a portion of the population with the generated offsprings for the next generation search. Common replacement approach in the literature may include:

– Replaces the entire population with the newly generated offsprings.
– Inspired by elitist strategy, some works proposed to copy the elite or a few elite parents to the next generation, and the rest of the population is replaced with the newly generated offspring.
– On the hand, the steady-state GA replaces just a few weak parents with new offsprings.
– Last but not least, the incremental GA replaces only one or two parents with the newly generated offsprings.

2.2 Memetic Algorithm

To enhance genetic algorithms by incorporating meme as a local search process, memetic search has been proposed in the literature as a form of population based search with lifetime learning as a separate process capable of local refinement for accelerating search and improving its quality.

Recent studies on MAs have demonstrated that they converge to high quality solutions more efficiently than their conventional counterparts [17–26] on many real world applications. To date, many dedicated MAs have been crafted to solve domain-specific problems more efficiently. In a recent special issue dedicated to MA research [27], several new design methodologies of memetic algorithms [27–33], such as specialized memetic algorithms designed for tackling the permutation flow shop scheduling [30], optimal control systems of permanent magnet synchronous motor [31], VLSI floor planning [34], quadratic assignment problem [35, 36], gene/feature selection [9], have been introduced. From a survey of the area, it is now well established that potential algorithmic improvement can be achieved by considering some important issues of MA [37–39]:

1. Local search frequency, hereby denoted as f_{il}: defines how often should individual local learning be applied. f_{il} can be represented as a percentage of the population, i.e., the percentage of individuals in the population that undergoes local learning, or the ratio of evolutionary to local search, i.e., in how many generations of global search should local learning be conducted. Alternatively, f_{il} can be replaced with the local search probability, P_{il}, which defines the probability at which each individual in the population should undergo local learning.
2. Local search intensity, t_{il}: defines how much computational budget should be allocated to each local learning process. t_{il} may be represented in terms of the number of objective function evaluations, or time budget.
3. Subset of solution undergoing local search, Ω_{il}: represents the subset of the solution population that undergoes local learning.
4. Local search method: should be carefully chosen from a given set of available local learning procedures, depending upon the problem to be solved.

2.3 Memetic Computation

Despite the success that has been achieved by incorporating meme as a local search process into evolutionary algorithms, such as the GA, the enhancement can often be very limited if the meme is inappropriately implemented (see the conditions listed in Sect. 2.2). For this reason, the learning of meme in GAs has been relegated, until now, to playing merely a complimentary role in the complete evolutionary cycle. Following the first definition of memes in Dawkins's book entitled "The selfish Gene" [6] as "the basic unit of cultural transmission via imitation", the true nature and potential merits of memes in enhancing the search in genetic algorithms remain yet to be fully exploited in the context of computational intelligence.

In a recent survey on the multi-facet views of memetic computation [5], it has been defined as a paradigm that uses the notion of meme(s) as units of information encoded in computational representations for addressing the ever increasing complexity and dynamic nature of problem-solving. This conceptualization thus unleashes a significant number of potentially rich meme-centric designs, operational models and algorithmic frameworks that could form the cornerstones of memetic computation as tools for effective genetic search.

In memetic computation, like gene in genetics, a meme is synonymous to memetic as being a building block of cultural know-how that is transmissible and replicable [4]. In the context of evolutionary computation, while genes are encoded as solutions of the problem of interest, meme shall denote the structured knowledge or latent pattern that can be transmitted between evolutionary searches on similar problem instances, and may lead to more focus and effective search, when appropriately harnessed.

3 Probablistic Memetic Algorithm

In this section, we present a recent study on formal memetic algorithm to balance between exploration and exploitation in search, via balancing between stochastic and deterministic operators. In memetic algorithm, while the issues summarized in Sect. 2.2 have been studied extensively in the literature, for example, Hart [40] and Ku [41] on the local search frequency, Land [42] on selecting appropriate individuals among the EA population that should undergo local search, Goldberg and Voessner [43] on local search intensity, Ong [17] and Kendall [44] on the selection of local search; it is worth noting that the works consider the design issues independently of each other. The recent work by Nguyen et al. [37] on the other hand proposed a formal theoretical probabilistic memetic framework (PrMF) that unifies the local search frequency, intensity and selection of solutions undergoing local search under a single theme.

3.1 Theoretical Upper Bound on Local Search

The theoretical upper bound on local search intensity in memetic algorithms, derived in [37], is given by:

$$t_{upper} = \frac{t_g}{n} \frac{ln(1 - p_2^{(k)})}{ln(1 - p_1^{(k)})} \tag{2}$$

where t_g denotes the function evaluations incurred in a generation and n is the population size. $p_1^{(k)}$ gives the probability that an individual, in generation k, lies infinitesimally close to the global optimum. $p_2^{(k)}$ is the probability that an individual in generation k lies within the basin of attraction of the global optimum. Based on Taylor series expansion and the assumption that the probabilities are small, the above equation was simplified to the following on configuring t_g to n:

$$t_{upper} = \frac{p_2^{(k)}}{p_1^{(k)}} \qquad (3)$$

By estimating $p_1^{(k)}$ and $p_2^{(k)}$ in the search process, this upper bound can be used to determine whether the current individual should undergo local search and/or the computational budget that should be allocated to the local search phase.

3.2 Estimation of Local Search Upper Bound

Figure 3 presents the depiction on the estimations of t_{upper} for continuous and combinatoric problems. In Fig. 3a, e denotes the acceptable tolerance for convergence to the global optimum and X is the current chromosome or solution. X_1 and X_2 are the nearest neighbors of X (based on simple Euclidean distance) selected from the database of solution vectors obtained by previous local search processes. Note that the local search traces for X_1 and X_2 are also depicted in the figure. In [37], the best solution A found in the neighborhood of X and the furthest search point B within the range of e from A are used to approximate $p_1 = \frac{|AB|}{volume\ of\ search\ space}$, and $p_2 = \frac{|AC|}{volume\ of\ search\ space}$, where $|\cdot|$ denotes the Euclidean distance. The upper bound for local search intensity, t_{upper}, on X is subsequently determined based on Eq. 3. On the other hand, the expected local search intensity, $t_{expected}$, required to reach the local optimum of X have been defined as the estimated average length of local search traces of X_1 and X_2 [37].

Further, since the definition of "nearest" neighbors is generally vague in the combinatorial context, in Fig. 3b, a single nearest neighbor of the current individual X has been identified as X_{nber} according to [45]. The local optimum reached, starting from X_{nber}, is labeled as X_A, while the individual solution found along the search trace before converging to X_A is labeled as X_B. Subsequently, the probabilities p_1 and p_2 have been approximated as $p_1 = \frac{Dis(X_B, X_A)}{volume\ of\ search\ space}$ and

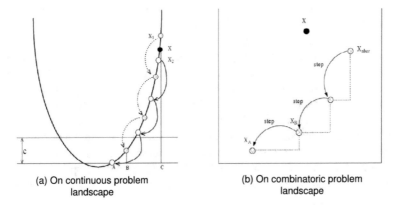

(a) On continuous problem landscape

(b) On combinatoric problem landscape

Fig. 3 Depiction on the estimations of t_{upper} for continuous and combinatoric problems

$p_2 = \frac{Dis(X, X_A)}{volume\ of\ search\ space}$. In [45], the $Dis(\cdot)$ measure between solutions has been defined as $|S_a \cup S_b| - |S_a \cap S_b|$, where S_a and S_b denote the two solutions of the problem of interests. The $|\cdot|$ operator gives the numbers of members (or cardinality) of a set. With these approximations, the upper bound t_{upper} can be derived with Eq. 3. Last but not the least, the expected local search intensity $t_{expected}$, which denotes the number of local search steps needed for X to reach X_A, has been approximated by the number of steps from X_{nber} to X_A in [45].

Algorithm 2 Outline of the probabilistic memetic framework

Begin:

Step 1: Identify the nearest neighbor of the given individual X from local search trace database

Step 2: Identify local optimum X_A of the nearest neighbors and solution X_B in the local searches

Step 3: Apply the probabilistic approach

    1. Estimate the upper bound for lifetime learning or local search

    2. Estimate the expected value for lifetime learning or local search

    3. if $t_{expected} \le t_{upper}$, local search will take place; otherwise, proceed with global exploration

End

We summarize the basic steps of the probabilistic memetic approach in Algorithm 2. Note that, in the context of combinatoric problems, the upper bound is considered only if the current solution X is sufficiently close to its nearest neighbor; otherwise the search will proceed with the local search process [45].

4 Meme-Centric Computing Paradigm for Search

Beyond the form of the local search or individual learning process, this section presets a recently proposed genetic search incorporating memes by treating meme as the build-block of problems that can be transmitted across similar problem instances for future effective problem-solving.

The essential ingredients of the presented meme-centric computing paradigm for search are depicted in Fig. 4. In the first step, memes of previously solved problem instances are captured via the meme learning process, which are stored in the meme pool or society of memes notated by **Ms**. For any given new unseen problem instance of the same domain, the meme selection process kicks in to identify a suitable meme from **Ms**, which is subsequently used to bias the solution generation during population initialization. Next, the conventional genetic search operators then proceed until the user-specified stopping condition is satisfied. The attained optimized solution together with the problem instance will also be archived for subsequent learning via the meme learning process.

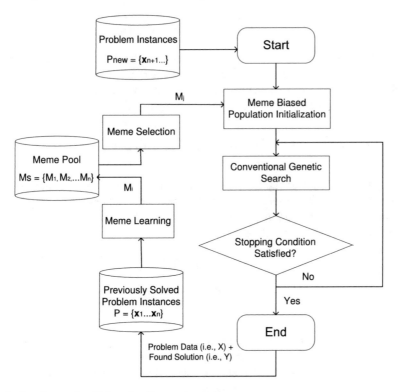

Fig. 4 Genetic search with learning from past experiences

In what follows, we present the specific realization of this meme-centric computing paradigm recently proposed in [11], for combinatorial routing problems.

4.1 Meme as Instruction for Task Assignment

In [11], meme has been defined as the instruction for mapping the distribution of customers of a given routing problem to its optimized solution. In particular, meme takes the form a positive semidefinite matrix that can scale the Euclidean distance among customers, which is given by:

$$d_M(\mathbf{x}_a, \mathbf{x}_b) = ||\mathbf{x}_a - \mathbf{x}_b||_M = \sqrt{(\mathbf{x}_a - \mathbf{x}_b)^T \mathbf{M}(\mathbf{x}_a - \mathbf{x}_b)}$$

where $\mathbf{x}_a = (x_{a1}, \ldots, x_{ap})^T$ and $\mathbf{x}_b = (x_{b1}, \ldots, x_{bp})^T$ denote two customers with location information. T denotes the transpose of a matrix or vector. Since $\mathbf{M}$ is positive semidefinite and, it can be decomposed as $\mathbf{M} = \mathbf{LL}^T$ by means of Singular Value decomposition. Substituting this decomposition into $d_M(\mathbf{x}_a, \mathbf{x}_b)$:

$$d_M(\mathbf{x}_a, \mathbf{x}_b) = \sqrt{(\mathbf{L}^T\mathbf{x}_a - \mathbf{L}^T\mathbf{x}_b)^T (\mathbf{L}^T\mathbf{x}_a - \mathbf{L}^T\mathbf{x}_b)} \qquad (4)$$

Thus, meme $\mathbf{M}$ gives the instruction of reinforcements to the customers representations in the problem space. Particularly, customers that are served by a common vehicle can be reinforced to be closer to one another while customers served by different vehicles can be kept further apart according to the optimized routing solution.

4.2 Learning and Selection of the Identified Meme

In [11], the learning and capturing of meme $\mathbf{M}$ from the optimized routing solution has been formulated as a maximization of the dependence between routing instance $\mathbf{X}$ and its optimized solution $\mathbf{y}*$ using the *Hilbert-Schmidt Independence Criterion* (HSIC) [46], which takes the form:

$$\max_{\mathbf{M}} \quad tr(\mathbf{XHYHX}^T\mathbf{M}) \qquad (5)$$
$$\text{s.t.} \quad \mathbf{M} \succeq 0$$
$$tr(\mathbf{M}^p) \leq B$$

where $tr(\cdot)$ denotes the trace operation of a matrix. $\mathbf{Y}$ is the kernel matrices for $\mathbf{y}*$. If task i and task j are served by the same vehicle, $\mathbf{Y}(i, j) = 1$, otherwise, $\mathbf{Y}(i, j) = -1$. $\mathbf{H} = \mathbf{I} - \frac{1}{n}\mathbf{11}'$ centers the data and the labels in the feature space, $\mathbf{I}$ denotes the identity matrix, n equals to the number of customers.

According to the Proposition in [47], $\mathbf{M}$ in Eq. 5 possesses a closed-form solution:

$$\mathbf{M} = \left(\frac{B}{trace(\mathbf{A}_+^{\frac{p}{p-1}})}\right)^{\frac{1}{p}} \mathbf{A}_+^{\frac{1}{p-1}} \qquad (6)$$

where $\mathbf{A}_+ = \mathbf{V}diag(\delta_+)\mathbf{V}^T$, and δ_+ is a vector with entries equal to $max(0, \delta[i])$. For $p = 1$, the optimal solution $\mathbf{M}$ can be expressed as closed-form solution:

$$\mathbf{M} = B\mathbf{A}_1 \qquad (7)$$

where $\mathbf{A}_1 = \mathbf{V}diag(\delta_1)\mathbf{V}^T$, and δ_1 is a vector with entries equal to $\frac{1}{\sum_{i:\delta_i=max(\delta)} 1}$ for all i that $\delta[i] = max(\delta)$, otherwise, the entries are zeros.

Further, to identify the suitable meme for enhanced genetic search on future unseen problems, the selection of meme in [11] has been formulated as:

$$\max_{\mu} \ tr(\mathbf{HX}^T\mathbf{M}_t\mathbf{XHY}) + \sum_{i=1}^{n} \mu_i S_i \tag{8}$$

$$\text{s.t. } \mathbf{M}_t = \sum_{i=1}^{n} \mu_i \mathbf{M}_i, \ \mu \in \mathcal{N}$$

$$\mathcal{N} = \{[\mu_i \dots \mu_n] \mid \sum_{i=1}^{n} \mu_i = 1, \mu_i \in \{0, 1\}, \forall i = 1 \dots n\}$$

$$\mathbf{M}_i \succeq 0$$

where S_i is the similarity measure between two given routing instance. $S_i = -(c1 * MMD_i + c2 * Dif_i)$, where MMD_i denotes the Maximum Mean Discrepancy (MMD) [48], which is used to compare the distribution discrepancy between two given instances by measuring the distance between their corresponding means. $MMD(D_s, D_t) = ||\frac{1}{n_s}\sum_{i=1}^{s}\phi(x_i^s) - \frac{1}{n_t}\sum_{i=1}^{t}\phi(x_i^t)||$, where $\phi(\cdot)$ maps the original input to a high dimensional space. In [11], the linear mapping has been adopted, i.e., $\phi(\mathbf{x}) = \mathbf{x}$. Dif_i denotes the difference in vehicle capacity for two given CARP instances, while $c1$ and $c2$ are the coefficients to balance the importance of differences in the tasks distribution and vehicle demand.

Taking the constraints into consideration, Eq. 8 can be reformulated as:

$$\max_{\mu \in \mathcal{N}} tr(\mathbf{HX}^T\mathbf{MXHY}) + \sum_{i=1}^{n} \mu_i S_i \tag{9}$$

$$\implies \max_{\mu \in \mathcal{N}} tr(\mathbf{HX}^T \sum_{i=1}^{n} \mu_i \mathbf{M}_i \mathbf{XHY}) + \sum_{i=1}^{n} \mu_i S_i \tag{10}$$

$$\implies \max_{\mu \in \mathcal{N}} \sum_{i=1}^{n} \mu_i (tr(\mathbf{HX}'\mathbf{M}_i\mathbf{XHY}) + S_i) \tag{11}$$

It is straightforward to see that, the matrix $\mathbf{M}_i$ that maximizes $tr(\mathbf{HX}^T\mathbf{M}_i\mathbf{XHY}) + S_i$ denotes the most suitable meme, and gives the corresponding $\mu_i = 1$. With two unknown variables (i.e., μ and $\mathbf{Y}$) in Eq. 11, an iterative method that keeps one variable fixed in a single generation, is adopted in [11] to solve this selection problem.

5 Case Studies

To verify the effectiveness of the presented probabilistic memetic algorithm (*PMA*) and meme-centric memetic computation paradigm (*MCP*), experimental studies conducted on the commonly used capacitated arc routing problem (*CARP*) benchmarks have been reported in [11, 45] and we summarize some of the findings and conclusion made in this section.

5.1 Capacitated Arc Routing Problem

The capacitated arc routing problem (CARP) was first proposed by Golden and Wong
[49] in 1981. It can be formally stated as follows: Given a connected undirected graph
$G = (V, E)$, where vertex set $V = \{v_i\}, i = 1 \ldots n, n$ is the number of vertices, edge
set $E = \{e_i\}, i = 1 \ldots m$ with m denoting the number of edges. Consider a demand
set $D = \{d(e_i)|e_i \in E\}$, where $d(e_i) > 0$ implies edge e_i requires servicing, a travel
cost vector $C_t = \{c_t(e_i)|e_i \in E\}$ with $c_t(e_i)$ representing the cost of traveling on
edge e_i, a service cost vector $C_s = \{c_s(e_i)|e_i \in E\}$ with $c_s(e_i)$ representing the cost
of servicing on edge e_i. A solution of CARP can be represented as a set of travel
circuits $S = \{C_i\}, i = 1 \ldots k$ which satisfies the following constraints:

1. Each travel circuit C_i, $i \in [1, k]$ must start and end at the depot node $v_d \in V$.
2. The total load of each travel circuit must be no more than the capacity W of each
   vehicle, $\sum_{\forall e_i \in C} d(e_i) \leq W$.
3. $\forall e_i \in E$ and $d(e_i) > 0$, there exists one and only one circuit $C_i \in S$ such that
   $e_i \in C_i$.

The cost of a travel circuit is then defined by the total service cost for all edges
that needed service together with the total travel cost of the remaining edges that
formed the circuit:

$$cost(C) = \sum_{e_i \in C_s} c_s(e_i) + \sum_{e_i \in C_t} c_t(e_i) \tag{12}$$

where C_s and C_t are edge sets that required servicing and those that do not, respectively. And the objective of CARP is then to find a valid solution S that minimizes
the total cost:

$$C_S = \sum_{\forall C_i \in S} cost(C_i) \tag{13}$$

The example of a CARP is illustrated in Fig. 5, with v_d representing the depot, full
line denoting edges that require servicing (otherwise known as tasks) and dashed lines
representing edges that do not require servicing. Each task is assigned a unique integer
number (e.g., 2 is assigned to the task from v_2 to v_1), the integer numbers enclosed
in brackets denoting the inversion of each task (i.e., direction of edge) accordingly.
In Fig. 5, three feasible solution circuits $C_1 = \{0, 4, 2, 0\}, C_2 = \{0, 5, 7, 0\}$, and $C_3 = \{0, 9, 11, 0\}$ can be observed, each composing of two tasks. A '0' index value is
assigned at the beginning and end of circuits to initialize each circuit to start and
end at the depot. According to Eqs. 12 and 13, the total cost of a feasible solution
$S = \{C_1, C_2, C_3\}$ is then obtained as sum of the service costs for all tasks and the
travel costs for all edges involved.

5.2 Experimental Configuration

Data Set: Accordingly to [11, 45], the well-established *egl* CARP benchmark has been used in this experimental study. The data set was generated by Eglese based on data obtained from the winter gritting application in Lancashire [50–52]. It consists of two series of data sets (i.e., "E" and "S" series) with a total of 24 instances. In particular, CARP instances in the "E" series have smaller number of vertices, tasks or edges than those in the "S" series. Thus, the problem structures of series "E" are deemed to be simpler as compared to series "S". The detailed properties of each *egl* instance are presented in Tables 1 and 2. "$|V|$", "$|E_R|$", "E" and "LB" denote the number of vertices, number of tasks, total number of edges and lower bound, of each problem instance, respectively.

Further, the state-of-the-art evolutionary search method recently proposed by Mei et al. [53] for solving CARP, denoted as *ILMA*, have been considered in [11, 45] as the baseline evolutionary solver in both of the probabilistic memetic algorithm and the meme-centric memetic computation paradigm.

Configuration of *PMA*: The same global and local search operators as well as other algorithmic configurations of *ILMA* [53] have been adopted. Algorithm 3 presents

Fig. 5 An example of CARP

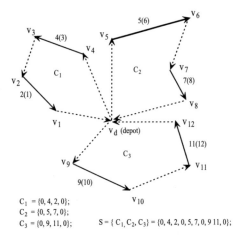

$C_1 = \{0, 4, 2, 0\};$
$C_2 = \{0, 5, 7, 0\};$
$C_3 = \{0, 9, 11, 0\};$
$S = \{C_1, C_2, C_3\} = \{0, 4, 2, 0, 5, 7, 0, 9\,11, 0\};$

Table 1 Properties of the *egl* "E" Series CARP benchmarks

Data set	"E" Series											
	E1A	E1B	E1C	E2A	E2B	E2C	E3A	E3B	E3C	E4A	E4B	E4C
V	77	77	77	77	77	77	77	77	77	77	77	77
E_r	51	51	51	72	72	72	87	87	87	98	98	98
E	98	98	98	98	98	98	98	98	98	98	98	98
LB	3548	4498	5566	5018	6305	8243	5898	7704	10,163	6048	8884	11,427

Table 2 Properties of the *egl* "S" Series CARP benchmarks. *Y*-axis: Fitness value, *X*-axis: Number of Fitness Evaluation or CPU Time in Seconds

Data set							"S" Series					
	S1A	S1B	S1C	S2A	S2B	S2C	S3A	S3B	S3C	S4A	S4B	S4C
V	140	140	140	140	140	140	140	140	140	140	140	140
E_r	75	75	75	147	147	147	159	159	159	190	190	190
E	190	190	190	190	190	190	190	190	190	190	190	190
LB	5018	6384	8493	9824	12,968	16,353	10,143	13,616	17,100	12,143	16,093	20,375

the *ILMA* equipped with the probabilistic framework. For *PMA*, the local search frequency has been fixed as *ILMA* before the first restart and subsequently adapted by the presented probabilistic approach. Further, as discussed in Sect. 3, the local search upper bound is considered when $Dis(X, X_{nber} \leq Dis(X_{nber}, X_A)$ (see Fig. 3).

Algorithm 3 Outline of *ILMA* equipped with the probabilistic framework

---

Begin:
Initialization: Generate the initial population
For the first restart
    While (the termination criteria are not met)
    Select two chromosomes from the current population
    Perform crossover operator to generate offspring
    Apply the local search process with tracking capability on the generated offspring with a
    certain probability
    Update the current population with the newly generated offspring
End For
For each of the subsequently restart
While (stopping conditions are not satisfied)
    Select two chromosomes from the current population
    Perform crossover operator to generate offspring
    Apply probabilistic memetic framework to the offspring
    Update the current population with the newly generated offspring
End While
End For
End

Configuration of *MCP*: To generate the pool of memes from past experiences, the CARP instance in *egl*, namely "E1A", "E1B", "E2A", "E3A", "E4A", "S1A", "S1B", "S2A", "S3A", "S4A" have been considered in [11] as the previously solved problem instances.

Subsequently, like in the traditional CARP, each task was represented by a corresponding *head vertex, tail vertex, travel cost* and *demand (service cost)*. To represent each task in the form of coordinates, in [11], the shortest distance matrix of the vertices were derived by means of the Dijkstra's algorithm [54], i.e., using the distances available between the vertices of a CARP. The coordinate features (i.e., locations) of each task were then approximated by means of multidimensional scaling [55]. In this manner, each task has been represented as a node in the form of coordinates. A CARP instance in this setting was thus represented by input vector **X** composing of the coordinate features of all tasks in the problem. Such a representation would allow the standard clustering approaches, such as the K-Means algorithm to be conducted on the CARP in task assignment. The label information of each task, i.e., in **Y** belonging to the CARP instance was then given by the optimized solution of baseline *ILMA*.

Further, according to [11], the MMD of Eq. 8 has been augmented with the demand of each task as additional problem feature. Coefficients $c1$ and $c2$ of Eq. 8 and parameters p and B for solving Eq. 5 were configured as 0.8, 0.2, 2 and 100, respectively.

Note that, the only difference between *MCP* and *ILMA* is the population initialization process. For *ILMA*, the initial population was generated according to [53], as a fusion of chromosomes generated from Augment_Merge [49], Path_Scanning [56], Ulusoy's Heuristic [57] and the simple random initialization procedures. For *MCP*, on the other hand, the initialization population was generated by meme biased K-Means clustering, which is learned from from past problem solving experiences.

5.3 *Results*

Table 3 summarizes the performance of *PMA* and *ILMA* on several metrics. "B. Cost", "Ave.Cost" and "Std.Dev" indicate the best result, averaged result and standard deviation obtained by the corresponding algorithms across 30 independent runs, respectively.

From Table 3, it can be observed that both *PMA* and *MCP* achieved improved performance when compared to the baseline *ILMA* in terms of "B. Cost" and "Ave.Cost" on all the CARP instances. With knowledge meme learned from past problem solving experiences and label information of past optimized CARP instances transferred to the current new unseen problem instance of interest, *MCP* obtained the best performance in term of "Ave.Cost" on 5 out of the first 7 "E" series CARP instance. It is worth noting that, *MCP* and *ILMA* share the same genetic solver and configuration, and only differ in the population initialization process. On the other hand, on the more complicated "S" series CARP instances, where the balance between global evolution and lifetime learning or local search becomes more important, the *PMA* which had a local search upper bound governing the global exploration and local exploitation process, demonstrated the best performance in terms of both "B. Cost" and "Ave.Cost" against the two comparisons.

6 Transferrable Memes in Evolutionary Bilevel Optimization

In this penultimate section, we provide a discussion on the potential application of the transferrable meme paradigm to a very important class of optimization problems which are ubiquitous in real world policy/decision making, but have only recently started garnering the interest of GA researchers. These problems, often termed as bilevel programs, are reincarnations of the classical Stackelberg duopoly in game theory. Such games reflect the rational choices of two competing decision makers (DMs) interacting sequentially, i.e. one DM (called the follower) makes a decision only after observing the decision of the other (called the leader). Clearly, the game is valid only if the objectives of a DM are such that they are directly affected by the choices of the other. It is strongly assumed, however, that the leader has complete

Table 3 Statistics of *ILMA*, *PMA* and *MCP* on *egl* CARP benchmarks

Data set	ILMA			PMA			MCP		
	B.Cost	Ave.Cost	Std.Dev	B.Cost	Ave.Cost	Std.Dev	B.Cost	Ave.Cost	Std.Dev
1. E1-C	5595	5602.1	8.9	5595	5597.1	5.7	5595	5596.3	6.9
2. E2-B	6317	6337.2	9.9	6317	6334.5	11.5	6317	6333.6	14.6
3. E2-C	8335	8350.6	29.2	8335	8338.2	12.2	8335	8347.2	22.7
4. E3-B	7777	7799.4	29.3	7775	7789.3	17.7	7775	7788.5	16.7
5. E3-C	10,292	10,325.4	28.8	10,292	10,313.7	38.9	10,292	10,310.1	16.9
6. E4-B	8998	9057.4	37.4	8988	9041.3	33.9	8988	9050.0	49.8
7. E4-C	11,609	11,714.6	78.8	11,594	11,658.1	55.6	11,542	11,648.1	65.2
8. S1-C	8519	8583.9	47.7	8518	8554.4	35.7	8518	8563.7	37.3
9. S2-B	13,190	13,278.3	63.8	13,153	13,250.3	75.1	13,157	13,252.2	60.4
10. S2-C	16,490	16,601.8	77.2	16,442	16,568.8	60.3	16,456	16,594.3	70.4
11. S3-B	13,784	13,910.3	60.6	13,711	13,843.0	67.9	13,757	13,885.1	78.8
12. S3-C	17,285	17,393.3	81.1	17,244	17,325.1	71.0	17,253	17,353.5	54.8
13. S4-B	16,394	16,529.6	66.7	16,331	16,449.6	63.4	16,358	16,506.1	68.5
14. S4-C	20,608	20,804.2	77.4	20,578	20,716.3	79.1	20,581	20,776.4	74.9

information about the possible choices and payoffs of the follower, so as not to be short-changed. On further considering the problem description it can be observed that the task is that of solving one optimization problem (that of the follower) nested within another (that of the leader). The formulation can mathematically be described as follows [58] (we consider a minimization problem here without any loss of generality):

$$Minimize \quad F(x_l, x_f) \tag{14}$$
$$s.\, t. \quad x_f = argmin\{f(x_f)given\ x_l\}$$
$$x_l = (x_1, \ldots, x_r); x_f = (x_{r+1}, \ldots, x_n)$$
$$x_i(L) \le x_i \le x_i(U), i = 1, \ldots, n.$$

In this description F is the objective function of the leader while f is the objective function of the follower. x_l and x_f are the choices/decisions of the leader and the follower, respectively. It is important to note here that the follower's problem is solved under a fixed x_l. Therefore x_f is considered a function of x_l.

Although it is not difficult to see the importance of bilevel programs in business and economic policy making, they are also very useful in engineering problems in which a process-design is considered feasible only if it ensures optimality of some sub-process (for a practical example of such a situation the reader is referred to [59] in which the composites manufacturing process is modelled as a game).

Despite the widespread applications of these models they have not been tackled satisfactorily in the GA literature. This is primarily because the nested nature of the problem makes it extremely computationally demanding, especially for a GA. As is well known, a standard GA typically takes a few thousand evaluations to find good solutions. However, in the case of bilevel programs each possible solution for the leader (a particular xl in Eq. 14) leads to a reparametrization or restructuring of the followers problem, thereby requiring a nested GA to be solved from scratch. This may cause the total number of function evaluations necessary for the complete problem to reach a few million. In cases where the cost associated with a single function evaluation is high, which is often the case with engineering problems, one must abandon all hopes of finding a good solution. It is to overcome this computational hurdle that the powers of the new meme-centric computing paradigm can be called upon. Recalling what has already been discussed, the meme can be perceived as a building-block of knowledge which essentially serves as a mapping between a problem instance and its optimal solution. Accordingly, during the initial stages of solving the leaders problem, if a meme learning procedure is carried out for the follower, this knowledge can be transferred directly to later generations under varying problem parameters and structures, without the need to solve a nested GA (at least not from scratch). In must be noted that the description of the meme itself may not have any resemblance to the one discussed in the previous section.

In order to fully explain the possible restructuring of the follower's problem, which is not appropriately captured through Eq. 14, we choose a common example from transportation research; in keeping with the spirit of the rest of this paper.

The "Toll Setting Problem" is one of finding the optimal set of tolls on different links (arcs) of an urban road network operated by an authority, so as to maximize some objective function. The authority here represents the leader in Eq. 14, while the users of the road are the followers who arrange themselves on the arcs of the network based on the policies set by the authority. Although the task at the lower level is to solve an equilibrium problem, i.e. each road user makes a route choice that is optimal given the choices of all the other network users, it is often modelled as a mathematical program through Beckmann's formulation [60]. While changes in network tolls can be seen as merely parametric changes to the follower's problem, a complete restructuring occurs when the arcs of the networks are to be cordoned off for the sake of maintenance, repair work or in response to road accidents. It is here again that the availability of a transferrable meme can play a critical role in enabling rapid policy making so as to reroute the network traffic in the most desirable manner.

Following the discussion in this section it can be seen that the concept of meme-centric computing or the amalgamation of transfer learning with GA, holds the key to a variety of pressing problems which are either intractable or too tedious for existing computational paradigms. Therefore, it provides a rich source of problems for researchers both in algorithm development as well as specific applications.

7   Conclusion

In this chapter, an introduction of genetic algorithms and their advances towards incorporating the learning of memes, has been presented. In particular, two research works that treat memes either as local search processes or as structured knowledge of problems, are introduced. In the first work, a probabilistic framework with a theoretical upper bound on local search intensity is described in order to balance the genetic search and local individual learning intensities while the search process is online. Further, a meme-centric computing paradigm for search is introduced, one that falls back on the basic notion of meme as the build-block of problems. The second work introduced learning and selection of memes in genetic search to transfer useful traits from past search experiences. Specific realization of this meme transfer learning on routing problems are provided. Case studies on the NP-hard capacitated arc routing benchmarks are carried out to show the efficacy of the two research works. Finally, a potential future application of the new memetic paradigm to a very important class of bilevel programming problems is discussed, so as to emphasize on the flexibility of the concept to encompass a wide variety of critical present-day challenges.

References

1. Goldberg, D.E.: Genetic Algorithms in Search, Optimization and Machine Learning, 1st edn. Addison-Wesley Longman Publishing Co., Inc., Boston (1989)
2. Holland, J.H.: Adaptation in Natural and Artificial Systems. University of Michigan Press, Ann Arbor (1975)
3. Bransford, J., Brown, A., Cocking, R. (eds) How People Learn: Brain, Mind, Experience, and School - Expanded Edition (2000)
4. Chen, X.S., Ong, Y.S., Lim, M.H., Tan, K.C.: A multi-facet survey on memetic computation. IEEE Trans. Evol. Comput. **5**, 591–607 (2011). (in Press)
5. Ong, Y.S., Lim, M.H., Chen, X.S.: Research frontier: memetic computation - past, present & future. IEEE Comput. Intell. Mag. **5**(2), 24–36 (2010)
6. Dawkins, R.: The Selfish Gene. Oxford University Press, Oxford (1976)
7. Neri, F., Cotta, C., Moscato, P.: Handbook of Memetic Algorithms. Studies in Computational Intelligence. Springer, Berlin (2011)
8. Tang, K., Mei, Y., Yao, X.: Memetic algorithm with extended neighborhood search for capacitated arc routing problems. IEEE Trans. Evol. Comput. **13**(5), 1151–1166 (2009)
9. Zhu, Z., Ong, Y.S., Dash, M.: Wrapper-filter feature selection algorithm using a memetic framework. IEEE Trans. Syst. Man Cybern. - Part B **37**(1), 70–76 (2007)
10. Meuth, R., Lim, M.H., Ong, Y.S., Wunsch, D.: A proposition on memes and meta-memes in computing for higher-order learning. Memetic Comput. **1**, 85–100 (2009)
11. Feng, L., Ong, Y.S., Tsang, I.W., Tan, A.H.: An evolutionary search paradigm that learns with past experiences. In: IEEE World Congress on Computational Intelligence, Congress on Evolutionary Computation (2012)
12. Louis, S.J., McDonnell, J.: Learning with case-injected genetic algorithms. IEEE Trans. Evol. Comput. **8**(4), 316–328 (2004)
13. Goldberg, D.E.: Genetic Algorithms in Search, Optimisation, and Machine Learning. Addison-Wesley, Reading, MA (1989)
14. Holland, J.H.: Adaptation in Natural and Artificial Systems. MIT Press, Cambridge, MA (1975)
15. Whitley, D.: A genetic algorithm tutorial. Stat. Comput. **4**, 65–85 (1994)
16. Thierens, D., Thierens, D.: Adaptive mutation rate control schemes in genetic algorithms. In: Proceedings of the 2002 IEEE World Congress on Computational Intelligence: Congress on Evolutionary Computation, pp. 980–985 (2002)
17. Ong, Y.S., Keane, A.J.: Meta-Lamarckian learning in memetic algorithms. IEEE Trans. Evol. Comput. **8**(2), 99–110 (2004)
18. Torn, A., Zilinskas, A.: Global Optimization. Lecture Notes in Computer Science, 350 (1989)
19. Houck, C., Joines, J., Kay, M.: Utilizing Lamarckian evolution and the Baldwin effect in hybrid genetic algorithms. NCSU-IE Technical Report 96-01, Meta-Heuristic Research and Applications Group, Department of Industrial Engineering, North Carolina State University (1996)
20. Vicini, A., Quagliarella, D.: Airfoil and wing design using hybrid optimization strategies. Am. Inst. Aeronaut. Astronaut. J. **37**(5), 634–641 (1999)
21. Ong, Y.S., Nair, P.B., Keane, A.J.: Evolutionary optimization of computationally expensive problems via surrogate modeling. Am. Inst. Aeronaut. Astronaut. J. **41**(4), 687–696 (2003)
22. Le, M.N., Ong, Y.S., Jin, Y.C., Sendhoff, B.: Lamarckian memetic algorithms: local optimum and connectivity structure analysis. Memetic Comput. **1**(3), 175–190 (2009)
23. Gong, M.G., Liu, C., Jiao, L.C., Cheng, G.: Hybrid immune algorithm with Lamarckian local search for multi-objective optimization. Memetic Comput. **2**(1), 47–67 (2009)
24. Ting, C.K., Ko, C.F., Huang, C.H.: Selecting survivors in genetic algorithm using tabu search strategies. Memetic Comput. **1**(3), 191–203 (2009)
25. Shinkyu, J., Hasegawa, S., Shimoyama, K., Obayashi, S.: Development and investigation of efficient GA/PSO-HYBRID algorithm applicable to real-world design optimization. IEEE Comput. Intell. Mag. **4**(3), 36–44 (2009)

26. Sattar, A., Seguier, R.: HMOAM: hybrid multi-objective genetic optimization for facial analysis by appearance model. Memetic Comput. **2**(1), 25–46 (2010)
27. Ong, Y.S., Krasnogor, N., Ishibuchi, H.: Special issue on memetic algorithm. IEEE Trans. Syst. Man Cybern. - Part B **37**(1), 2–5 (2007)
28. Lim, M.H., Xu, Y.L.: Application of hybrid genetic algorithm in supply chain management. Int. J. Comput. Syst. Sign. **6**(1) (2005)
29. Smith, J.E.: Co-evolving memetic algorithms: a review and progress report. IEEE Trans. Syst. Man Cybern. - Part B **37**(1), 6–17 (2007)
30. Liu, B., Wang, L., Jin, Y.H.: An effective PSO-based memetic algorithm for flow shop scheduling. IEEE Trans. Syst. Man Cybern. - Part B **37**(1), 18–27 (2007)
31. Caponio, A., Cascella, G.L., Neri, F., Salvatore, N., Sumne, M.: A fast adaptive memetic algorithm for online and offline control design of PMSM drives. IEEE Trans. Syst. Man Cybern. - Part B **37**(1), 28–41 (2007)
32. Liu, D., Tan, K.C., Goh, C.K., Ho, W.K.: A multiobjective memetic algorithm based on particle swarm optimization. IEEE Trans. Syst. Man Cybern. - Part B **37**(1), 42–50 (2007)
33. Hasan, S.M.K., Sarker, R., Essam, D., Cornforth, D.: Memetic algorithms for solving job-shop scheduling problems. Memetic Comput. **1**(1), 69–83 (2008)
34. Tang, M., Yao, X.: A memetic algorithm for VLSI floorplanning. IEEE Trans. Syst. Man Cybern. - Part B **37**(1), 62–69 (2007)
35. Tang, J., Lim, M.H., Ong, Y.S.: Parallel memetic algorithm with delective local search for large scale quadratic assignment problems. Int. J. Innovative Comput. Inf. Control **2**(6), 1399–1416 (2006)
36. Tang, J., Lim, M.H., Ong, Y.S.: Diversity-adaptive parallel memetic algorithm for solving large scale combinatorial optimization problems. Soft Comput. J. **11**(9), 873–888 (2007)
37. Nguyen, Q.H., Ong, Y.S., Lim, M.H.: A probabilistic memetic framework. IEEE Trans. Evol. Comput. **13**(3), 604–623 (2009)
38. Ong, Y.S., Lim, M.H., Zhu, N., Wong, K.W.: Classification of adaptive memetic algorithms: a comparative study. IEEE Trans. Syst. Man Cybern. B Cybern. **36**(1), 141–152 (2006)
39. Lozano, M., Herrera, F., Krasnogor, N., Molina, D.: Real-coded memetic algorithms with crossover hill-climbing. Evol. Comput. **12**(3), 273–302 (2004)
40. Hart, W.E.: Adaptive global optimization with local search. Ph.D. thesis, University of California, San Diego (1994)
41. Ku, K.W.C., Mak, M.W., Siu, W.C.: A study of the Lamarckian evolution of recurrent neural networks. IEEE Trans. Evol. Comput. **4**(1), 31–42 (2000)
42. Land, M.W.S.: Evolutionary algorithms with local search for combinatorial optimization. Ph.D. Thesis, University of California, San Diego (1998)
43. Goldberg, D.E., Voessner, S.: Optimizing global-local search hybrids. Genet. Evol. Comput. Conf. **1**, 220–228 (1999)
44. Kendall, G., Cowling, P., Soubeiga, E.: Choice function and random hyperheuristics. In: Fourth Asia-Pacific Conference on Simulated Evolution and Learning, pp. 667–671 (2002)
45. Feng, L., Ong, Y.S., Nguyen, Q.H., Tan, A.-H.: Towards probabilistic memetic algorithm: an initial study on capacitated arc routing problem. In: IEEE Congress on Evolutionary Computation, pp. 18–23 (2010)
46. Gretton, A., Bousquet, O., Smola, A., Schölkopf, B.: Measuring statistical dependence with hilbert-schmidt norms. In: Proceedings Algorithmic Learning Theory, pp. 63–77 (2005)
47. Zhuang, J., Tsang, I., Hoi, S.C.H.: A family of simple non-parametric kernel learning algorithms. J. Mach. Learn. Res. (JMLR) **12**, 1313–1347 (2011)
48. Borgwardt, K.M., Gretton, A., Rasch, M.J., Kriegel, H.P., Schölkopf, B., Smola, A.J.: Integrating structured biological data by kernel maximum mean discrepancy. In: Proceedings of the 14th International Conference on Intelligent Systems for Molecular Biology, pp. 49–57 (2006)
49. Golden, B.L., Wong, R.T.: Capacitated arc routing problems. Networks **11**(3), 305–315 (1981)
50. Eglese, R.W.: Routing winter gritting vehicles. Discrete Appl. Math. **48**(3), 231C–244 (1994)
51. Eglese, R.W., Li, L.Y.O.: A tabu search based heuristic for arc routing with a capacity constraint and time deadline. In: Osman, I.H., Kelly, J.P. (eds.) Metaheuristics: Theory and Applications, pp. 633C–650. Kluwer Academic Publishers, Boston (1996)

52. Li, L.Y.O., Eglese, R.W.: An interactive algorithm for vehicle routing for winter-gritting. J. Oper. Res. Soc. **47**(2), 217C–228 (1996)
53. Mei, Y., Tang, K., Yao, X.: Improved memetic algorithm for capacitated arc routing problem. In: IEEE Congress on Evolutionary Computation, pp. 1699–1706 (2009)
54. Dijkstra, E.W.: A note on two problems in connection with graphs. Numer. Math. **1**, 269C–271 (1959)
55. Borg, I., Groenen, P.J.F.: Modern Multidimensional Scaling: Theory and Applications. Springer, Berlin (2005)
56. Golden, B.L., DeArmon, J.S., Baker, E.K.: Computational experiments with algorithms for a class of routing problems. Comput. Oper. Res. **10**(1), 47–59 (1983)
57. Ulusoy, G.: The fleet size and mix problem for capacitated arc routing. Eur. J. Oper. Res. **22**(3), 329–337 (1985)
58. Sinha, A., Malo, P., Deb, K.: Test problem construction for single-objective bilevel optimization. Evol. Comput. J. (2014)
59. Gupta, A., Kelly, P.A., Ehrgott, M., Bickerton, S.: A surrogate model based evolutionary game-theoretic approach for optimizing non-isothermal compression RTM processes. Compos. Sci. Technol. **84**, 92–100 (2013)
60. Larsson, T., Lindberg, P.O., Patriksson, M., Rydergren, C.: On traffic equilibrium models with a nonlinear time/money relation. In: Patriksson, M., Labbe, M. (eds.) Transportation Planning: State of the Art, pp. 19–31. Kluwer Academic Publishers, Dordrecht (2002)

Constrained Multi-objective Evolutionary Algorithm

Kalyanmoy Deb

Abstract Multi-objective optimization problems are common in practice. In practical problems, constraints are also inevitable. The population approach and implicit parallel search ability of evolutionary algorithms have made them popular and useful in finding multiple trade-off Pareto-optimal solutions in multi-objective optimization problems since the past two decades. In this chapter, we discuss evolutionary multi-objective optimization (EMO) algorithms that are specifically designed for handling constraints. Numerical test problems involving constraints and some constrained engineering design problems which are often used in the EMO literature are discussed next. The chapter is concluded with a number of future directions in constrained multi-objective optimization area.

Keywords Multi-objective optimization · Constrained optimization
Evolutionary algorithms · Pareto-optimal solution

1 Introduction

Most practical optimization problems cannot be formulated without a constraint. Constraints restrict some relationships which the variables corresponding to a feasible solution must follow. Most occasions, the constraints impose restrictions for a physical realization of a solution from space or fabrication considerations, for policy regulations involving environment and society, for functional requirements, for resource limitations, etc. Thus, an optimization of a practical problem without considering any constraint may not lead to an acceptable solution.

Parts of this chapter is excerpted from author's 2001 Wiley book [8] and his other publications.

K. Deb (✉)
Computational Optimization and Innovation (COIN) Laboratory, Department of Electrical and
Computer Engineering, Michigan State University, East Lansing, MI 48864, USA
e-mail: kdeb@egr.msu.edu
URL:http://www.egr.msu.edu/ kdeb

© Springer International Publishing AG, part of Springer Nature 2019   85
J. C. Bansal et al. (eds.), *Evolutionary and Swarm Intelligence*
Algorithms, Studies in Computational Intelligence 779,
https://doi.org/10.1007/978-3-319-91341-4_6

Despite the importance of constraints in optimization, many evolutionary optimization (EO) researchers suggest algorithms without indicating any viable way of handling constraints. The reason for such practices lie in the fundamental 'disciplinary practices' that most evolutionary algorithm researchers are familiar with. In computer science related optimization tasks, most problems are unconstrained and researchers consider constraint handling as an add-on secondary activity. Many efficient unconstrained EO algorithms have never been complemented with any constrained handling procedure. Although initial suggestions of multi-objective test problems, such ZDT series [37] and WFG test problems [21], were unconstraints with box constraints alone, these problems were not extended to include constraints. On the contrary, every classical optimization algorithm carefully suggest not only a constraint handling procedure, it also indicates clearly a termination criterion and provide some mathematical or otherwise reasons for every step of the suggested algorithm. In this author's view, EO researchers should pay more attention to constraint handling aspect of their proposed algorithms and also provide a reasonable criterion for terminating a simulation run.

This chapter focuses on constrained handling methods for multi- and many-objective evolutionary optimization, as evolutionary methods, since their popularity since early eighties, have been increasingly used for solving these problems. The reason for EO's popularity in solving multi-objective optimization problems is their ability to (i) find multiple trade-off solutions in a single simulation, (ii) constitute an efficient parallel search due to their population approach, and (iii) be flexible in changing their operators to suit to solve different problems without much change in their structure. The research and application in evolutionary multi-objective optimization (EMO) started in early nineties and have been quickly adopted by industries and various fields. After showing their usefulness in two and three objectives, they have been extended to solve four and more objective problems, as the practice demands such problem solving as well. It turns out that higher dimensional problems require an explicit guidance of some sort to solve these so-called many-objective optimization problems efficiently. Recently proposed many-objective optimization algorithms have shown to solve as large as 15 to 20-objective optimization problems. With all around activities in multi- and many-objective optimization problems, there is a growing need for algorithms for handling constraints, a matter which we discuss in this chapter.

In the remainder of this chapter, we outline a brief introduction to EMO principle for solving multi-objective optimization problems in Sect. 2. In Sect. 3, we present earlier and current constrained handling methods. Thereafter, in Sect. 4, we discuss constrained test problems for multi-objective optimization. Finally, conclusions of this chapter are drawn in Sect. 6.

2 Evolutionary Multi-objective Optimization (EMO)

A multi-objective optimization problem involves multiple conflicting objective functions which are to be either minimized or maximized subject to a number of constraints and variable bounds:

$$\left.\begin{array}{ll} \text{Minimize/Maximize } f_m(\mathbf{x}), & m = 1, 2, \ldots, M; \\ \text{subject to } g_j(\mathbf{x}) \geq 0, & j = 1, 2, \ldots, J; \\ h_k(\mathbf{x}) = 0, & k = 1, 2, \ldots, K; \\ x_i^{(L)} \leq x_i \leq x_i^{(U)}, & i = 1, 2, \ldots, n. \end{array}\right\} \quad (1)$$

A solution $\mathbf{x} \in \mathbf{R}^n$ is a vector of n decision variables: $\mathbf{x} = (x_1, x_2, \ldots, x_n)^T$. The solutions satisfying all constraints and variable bounds constitute a S in the decision variable space $\mathbf{R}^n$. The corresponding set of points in the objective space is called the Z.

In a multi-objective optimization problem involving conflicting objectives, there is more than one optimal solutions, known as *Pareto-optimal* solutions. The corresponding objective vectors are known *efficient points* or *non-inferior points*. It is obvious that every Pareto-optimal solution must be a feasible solution. The definition of a Pareto-optimal solution is related to the concept of domination, as follows [8, 27]:

Definition 11 A solution $\mathbf{x}^{(1)}$ is said to dominate another solution $\mathbf{x}^{(2)}$, if both the following conditions are true:

1. The solution $\mathbf{x}^{(1)}$ is no worse than $\mathbf{x}^{(2)}$ in all objectives. Thus, the solutions are compared based on their objective function values (or location of the corresponding points ($\mathbf{z}^{(1)}$ and $\mathbf{z}^{(2)}$) in the objective function set Z).
2. The solution $\mathbf{x}^{(1)}$ is strictly better than $\mathbf{x}^{(2)}$ in at least one objective.

A Pareto-optimal solution is such a solution that does not dominated by any feasible solution in the search space. Interestingly, there may exist more than Pareto-optimal solutions, each of which does not get dominated by any feasible solution, including other Pareto-optimal solutions and the set of all Pareto-optimal solutions are said to be *non-dominated* to each other, meaning that there exists a trade-off between any two Pareto-optimal solutions. That is, to move from one Pareto-optimal solution to another, there is always a gain in at least one of the objectives and this gain comes only from a sacrifice from at least of the other objectives. For a two-objective problem, the scenario is clear. Between two Pareto-optimal solutions, there is a gain in one of the objectives and there is a loss in the other objective. With more than two objectives, there are many ways trade-off can occur, hence there exist different levels of trade-off which could be of interest to users. Most EMO algorithms use the above domination principle to drive their search towards Pareto-optimal set, but there exist rigorous theoretical optimality conditions for identifying a Pareto-optimal solution [16, 27]. For differentiable problems, a solution $\mathbf{x}^*$ satisfying the following condition.

Theorem 1 (Fritz-John necessary condition) *A necessary condition for* $\mathbf{x}^*$ *to be Pareto-optimal is that there exist vectors* $\lambda \geq \mathbf{0}$ *and* $\mathbf{u} \geq \mathbf{0}$ *(where* $\lambda \in \mathbb{R}^M$*,* $\mathbf{u} \in \mathbb{R}^J$ *and* $\lambda, \mathbf{u} \neq \mathbf{0}$*) such that the following conditions are true:*

1. $\sum_{m=1}^{M} \lambda_m \nabla f_m(\mathbf{x}^*) - \sum_{j=1}^{J} u_j \nabla g_j(\mathbf{x}^*) = \mathbf{0}$*, and*
2. $u_j g_j(\mathbf{x}^*) = 0$ *for all* $j = 1, 2, \ldots, J$*.*

For a proof, readers may refer to Cunha and Polak [4]. Those readers familiar with the Kuhn-Tucker necessary conditions for single-objective optimization will immediately recognize the similarity between the above conditions and that of the single-objective optimization. The difference is in the inclusion of a λ-vector with the gradient vector of the objectives.

The following theorem offers sufficient conditions for a solution to be Pareto-optimal for convex functions.

Theorem 2 (Karush-Kuhn-Tucker sufficient condition for Pareto-optimality) *Let the objective functions be convex and the constraint functions of the problem shown in Eq. (1) be non-convex. Let the objective and constraint functions be continuously differentiable at a feasible solution* $\mathbf{x}^*$*. A sufficient condition for* $\mathbf{x}^*$ *to be Pareto-optimal is that there exist vectors* $\lambda > \mathbf{0}$ *and* $\mathbf{u} \geq \mathbf{0}$ *(where* $\lambda \in \mathbb{R}^M$ *and* $\mathbf{u} \in \mathbb{R}^J$*) such that the following equations are true:*

1. $\sum_{m=1}^{M} \lambda_m \nabla f_i(\mathbf{x}^*) - \sum_{j=1}^{J} u_j \nabla g_j(\mathbf{x}^*) = \mathbf{0}$*, and*
2. $u_j g_j(\mathbf{x}^*) = 0$ *for all* $j = 1, 2, \ldots, J$*.*

For a proof, see [27]. If the objective functions and constraints are not convex, the above theorem does not hold. However, for pseudo-convex and non-differentiable problems, different necessary and sufficient conditions exist [1].

2.1 EMO Algorithms

It is clear from the above description that a multi-objective optimization problem (MOOP) gives rise to a number of Pareto-optimal solutions. Although one of the Pareto-optimal solutions needs to be chosen eventually, it is argued [8] that the a posteriori approach of finding a representative set of Pareto-optimal solutions helps to make a more informed and judicious decision-making than the a priori approaches in which objectives are scalarized to a single objective prior to the optimization process.

Classical point-by-point optimization algorithms use the so-called *generative* method in which a parameterized single-objective optimization problem is solved multiple times, each with a different parameter value so as to generate a representative set of Pareto-optimal solutions. Since an evolutionary algorithm (EA) works with a population of solutions in every iteration, an EA can be applied only once with N population members so to find at most N Pareto-optimal solutions in a single simulation run. It has been argued and demonstrated elsewhere [31] that the evolutionary approach is faster and more efficient than the classical generative approach due to EA's implicit parallel processing ability. Simply stated, a lone population member lying close to a Pareto-optimal solution can participate with other population members through EA's recombination operator to bring them closer to other Pareto-optimal solutions.

The research and application of EMO started in the beginning of nineties with the introduction of three competent EMO algorithms (multi-objective genetic algorithm (MOGA) [17], niched Pareto GA (NPGA) [20], and non-dominated sorting GA (NSGA) [32]). Each of these early methods had at least one niching parameter which were needed to be tuned for every problem and lacked any elite preservation mechanism, which was found to be useful for single-objective optimization problems. However, the second-generation methods such as elitist non-dominated sorting GA (NSGA-II) [9], strength Pareto EA (SPEA) [38], and Pareto-archived evolution strategy (PAES) [24] suggested around the year 2000 eliminated both the above shortcomings and became state-of-the-art EMO algorithms. Although these EMO methods made the task of multi-objective optimization widespread in various scientific, engineering, and commercial areas, they were limited to handle two or three-objective optimization problems [23]. More recent EMO methodologies (MOEA/D [35] and NSGA-III [13, 22]) are capable of handling many-objective optimization problems involving 10–15 objectives.

3 Constrained EMO Algorithms

Despite the systematic development of EMO algorithms for solving unconstrained MOOP problems and despite the importance of solving constrained problems from practice, EMO researchers have not paid much attention in developing efficient constrained handling methods. In this section, we discuss some early methods and then discuss a couple of methods, one of which has so far remained a standard way of handling constraints in EMO literature.

3.1 Penalty Function Method

One simple approach for handling constraints is to penalize infeasible solutions by adding a penalty term to the objective function proportional to constraint violation [6, 30]. However, in a MOOP problem, one independent penalty term must be added for each objective function.

Before the constraint violation is calculated, all constraints are normalized. A constraint $g_j(\mathbf{x}) \geq b_j$ is converted into its normalized form, as follows:

$$\bar{g}_j(\mathbf{x}) \equiv g_j(\mathbf{x})/b_j - 1 \geq 0. \qquad (2)$$

Since each constraint is now normalized, they all can be added together to compute an overall constraint violation:

$$\Omega(\mathbf{x}) = \sum_{j=1}^{J} \langle \bar{g}_j(\mathbf{x}) \rangle, \tag{3}$$

where $\langle \alpha \rangle = -\alpha$, if $\alpha < 0$; zero, otherwise. The constraints can also be normalized using the population minimum and maximum g_j values before computing the constraint violation.

This constraint violation is then multiplied with a penalty parameter R_m and the product is added to each of the objective function values:

$$F_m(\mathbf{x}^{(i)}) = f_m(\mathbf{x}^{(i)}) + R_m \Omega(\mathbf{x}^{(i)}). \tag{4}$$

The functions F_m takes into account the constraint violations. For a feasible solution, the corresponding Ω term is zero and F_m becomes equal to the original objective function f_m. However, for an infeasible solution, $F_m > f_m$, achieved by adding a penalty corresponding to total constraint violation. The penalty parameter R_m is used to make both of the terms on the right side of the above equation to have the same order of magnitude. Since the original objective functions could be of different magnitudes, the penalty parameter must also vary from one objective function to another. Fixing M appropriate penalty parameter is the drawback of the penalty function approach. However, a number of static and dynamic strategies to update the penalty parameter were suggested in the single-objective GA literature [19, 25, 26]. Any of these techniques can be used for MOOP as usual. However, most studies in multi-objective evolutionary optimization use carefully chosen static values of R_m [7, 32]. Once the penalized function (Eq. 4) is formed, any of the unconstrained EMO methods can be used to optimize all F_m. Since all penalized functions will be minimized, an EMO should move into the feasible region and finally approach the Pareto-optimal set.

3.2 Deb's Parameter-Less Method

Deb's parameter-less constraint handling method uses the binary , where two solutions are picked from the population and the better solution is chosen. In the presence of constraints, each solution can be either feasible or infeasible. Thus, there may be at most three situations: (i) both solutions are feasible, (ii) one is feasible and other is not, and (iii) both are infeasible. For single-objective optimization, we used a simple rule for each case:

Case (i) Choose the solution with the better objective function value.
Case (ii) Choose the feasible solution.
Case (iii) Choose the solution with smaller overall constraint violation.

When both solutions are feasible [Case (i)], we check if they belong to separate non-dominated fronts. In such an event, we choose the one that belongs to the better non-dominated front. If they belong to the same non-dominated front, we check the diversity associated with each solution to resolve the tie. Since maintaining diversity is one of the two goals in multi-objective optimization, we choose the one which belongs to the least crowded region in that non-dominated set.

We define the following *constrain-domination* condition for any two solutions $x^{(i)}$ and $x^{(j)}$ for implementing the above concept:

Definition 12 A solution $x^{(i)}$ is said to 'constrain-dominate' a solution $x^{(j)}$ (or $x^{(i)} \preceq_c x^{(j)}$), if any of the following conditions are true:

1. Solution $x^{(i)}$ is feasible and solution $x^{(j)}$ is not.
2. Solutions $x^{(i)}$ and $x^{(j)}$ are both infeasible, but solution $x^{(i)}$ has a smaller *constraint violation*.
3. Solutions $x^{(i)}$ and $x^{(j)}$ are feasible and solution $x^{(i)}$ dominates solution $x^{(j)}$ in the usual sense (see Definition 11 above).

The above definition of constrain-domination will allow a similar non-dominated classification in the feasible region, but will classify the infeasible solutions according to their constraint violation values. Usually, each infeasible solution will belong to a different non-constrain-dominated front in the order of their constraint violation values, except when more than one solutions have an identical constraint violation.

The second step above requires the computation of an overall constraint violation, a matter which has prompted many researchers to define constraint violation differently. In Deb's approach, each constraint function is normalized using Eq. (3). By using the above definition of constrain-domination, we now define a generic constrained tournament selection operator as follows.

Definition 13 Given two solutions $x^{(i)}$ and $x^{(j)}$, choose solution $x^{(i)}$ if any of the following conditions are true:

1. Solution $x^{(i)}$ belongs to a better non-constrain-dominated set.
2. Solutions $x^{(i)}$ and $x^{(j)}$ belong to the same non-constrain-dominated set, but solution $x^{(i)}$ resides in a less crowded region based on a niched-distance measure.

The niched-distance measure refers to the measure of density of the solutions in the neighborhood of a solution. The niched-distance can be computed by using various metrics, as follows.

Niche count metric. Calculate the niche count nc_i and nc_j with solutions of the non-constrain-dominated set by using the sharing function method [11]. Since the niche count of a solution gives an idea of the number of crowded solutions and their relative distances from the given solution, a smaller value of the niche count means fewer solutions in the neighborhood. Thus, if this metric is used, the solution with the smaller niche count must be chosen. Unfortunately, this metric requires a parameter σ_{share}, due to which no further studies had been pursued.

Head count metric. Instead of computing the niche count, the total number of solutions (the head count) in the σ_{share}-neighborhood can be counted for each solution $\mathbf{x}^{(i)}$ and $\mathbf{x}^{(j)}$. Using this metric, choose the solution with the smaller head count. This metric may lead to ties, which can be avoided by using the niche count metric. This is because the niche count metric better quantifies the crowding of solutions. As for the niche count metric, the neighborhood checking can be performed either in the decision variable space or in the objective space. This metric also requires a niching parameter which must be set right for an arbitrary problem, but if objective space niching is performed, the niching parameter can be computed using a procedure suggested elsewhere [17].

Crowding distance metric. This metric is used in the NSGA-II. This metric estimates half of the perimeter of the maximum hypercube which can be allowed around a solution without including any other solution from the same obtained non-constrain-dominated front inside the hypercube. A large crowding distance for a solution implies that the solution is less crowded. The extreme solutions are assigned an infinite crowding distance. This measure requires $O(N_p \log N_p)$ computations, because of the involvement of the sorting of solutions in all M objectives. This crowding distance can be computed in the decision variable space as well.

3.3   Fonseca and Fleming's Method

Fonseca and Fleming's [18] method suggested a generic constrained satisfaction and handling method. For constrained optimization problems, the idea is similar in principle to Deb's parameter-less approach described above. The only difference is in the way domination is defined for the infeasible solutions. In the above definition, an infeasible solution having a larger overall constraint-violation is classified as a member of a larger non-constrain-domination level. On the other hand, in [18], infeasible solutions violating different constraints are classified as members of the same non-constrain-dominated front. Drechsler [15] also proposed a more detailed ordering approach for constraint violation. In these approaches one infeasible solution violating a constraint g_j marginally will be placed in the same non-constrain-dominated level with another solution violating a different constraint to a large extent but not violating g_j. This may cause an algorithm to wander in the infeasible search region for more generations before reaching the feasible region through some constraint boundaries. Moreover, since these approaches require domination checks to be performed with the constraint-violation values, they are supposedly computationally more expensive. However, a careful study is needed to investigate if the added complexity introduced by performing a non-domination check over the simple procedure described earlier in this section is beneficial to certain problems. Despite suggestion of these methods for more than two decades ago, such a study is left unperformed.

Besides simply adding the constraint violations together (in Deb's method), Binh and Korn [2] suggested a different constraint violation measure:

$$C(\mathbf{x}) = \left(\sum_{j=1}^{J} [c_j(\mathbf{x})]^p \right)^{\frac{1}{p}}, \tag{5}$$

where $p > 0$ and $c_j(\mathbf{x}) = |\min(0, g_j(\mathbf{x}))|$ for all j. In Deb's approach, $p = 1$ is used. In addition, Binh and Korn's constrained (MOBES) approach used a careful comparison of the feasible and infeasible solutions based on a diversity-preservation concept. These investigators preferred an infeasible solution over a feasible solution, if the infeasible solution resides in a less crowded region in the search space. By requiring a user-defined , was able to find a widely distributed set of non-constrain-dominated solutions in two test problems. Further studies are needed to justify the added complexities of the MOBES compared to the simplistic parameter-less approach of Deb [8].

Ray et al. [29] suggested a more elaborate constraint handling technique, where the constraint violations of all constraints are not simply added together; instead, a non-domination check of the constraint violations is made. Since this method has not been used much in EMO applications, we do not discuss this rather complex method here. Interested readers may refer to the original study.

4 Constrained Multi-objective Test Problems

A giant leap towards the development of EMO field came from the suggestion of test problems, as they allowed EMO algorithms to be compared against each other and provided the much-needed information about the weaker aspects of an EMO algorithm. The implication of test problems is also the same for constrained EMO methods. In the following subsections, we present the constrained multi-objective test problems that were used extensively for testing the performance of constrained EMO algorithms.

Constraints may cause hindrance for an EMO to converge to the true Pareto-optimal region and may also cause difficulty in maintaining a diverse set of Pareto-optimal solutions. It is intuitive that the success of an EMO in tackling both of these hindrances will largely depend on the constraint-handling technique used.

4.1 Specific Two-Objective Problems

Veldhuizen [34] has collected a number of constrained test problems used by several researchers. It is evident from the survey that most constrained test problems used only two to three variables and constraints are not sufficiently nonlinear. In the following, we present three such test problems commonly used by EMO researchers.

Test Problem BNH Binh and Korn [2] used the following two-variable constrained problem:

$$\text{Minimize } f_1(\mathbf{x}) = 4x_1^2 + 4x_2^2,$$
$$\text{Minimize } f_2(\mathbf{x}) = (x_1 - 5)^2 + (x_2 - 5)^2,$$
$$\text{subject to } C_1(\mathbf{x}) \equiv (x_1 - 5)^2 + x_2^2 \le 25,$$
$$C_2(\mathbf{x}) \equiv (x_1 - 8)^2 + (x_2 + 3)^2 \ge 7.7,$$
$$0 \le x_1 \le 5,$$
$$0 \le x_2 \le 3.$$

(6)

Figure 1 shows the feasible decision variable space (the shaded region enclosed by OBDEO). Both constraints (C_1, C_2) are also shown on the figure. It is interesting to note that the second constraint C_2 is redundant and does not make any part of the bounded space infeasible. On the other hand, constraint C_1 makes the region $OABO$ infeasible. Figure 2 shows the corresponding objective space. Since constraint C_2 does not affect the feasible region, we do not show this constraint in the objective

Fig. 1 Decision variable space for the test problem BNH

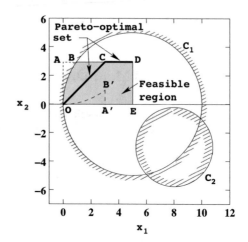

Fig. 2 Objective space for the test problem BNH

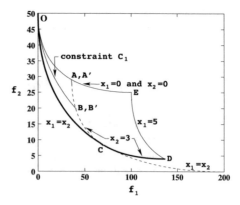

space. The constraint boundary of C_1 is also shown in this figure. There exists an equivalent region $OA'B'O$ in the decision variable space which maps to the same region $OABO$ in the objective space. Thus, although some portion of the objective space is eliminated because of the constraint C_1, no solution of the unconstrained bounded search space gets eliminated in the objective space. Thus, the presence of the first constraint is also not critical in this problem. The only difficulty that this constraint creates is that it reduces the density of solutions in the region $OABO$ in the objective space. The shape and continuity of the Pareto-optimal set is unchanged by the inclusion of both constraints.

The Pareto-optimal solutions [8] are constituted by solutions $x_1^* = x_2^* \in [0, 3]$ (region OC) and $x_1^* \in [3, 5]$, $x_2^* = 3$ (region CD). These solutions are marked by using bold continuous curves. The addition of both constraints in the problem does not make any solution in the unconstrained Pareto-optimal front infeasible. Thus, constraints in this problem may not introduce any additional difficulty in solving the problem.

Test Problem OSY Osyczka and Kundu [28] used the following six-variable test problem:

$$\text{Minimize } f_1(\mathbf{x}) = -\left[25(x_1 - 2)^2 + (x_2 - 2)^2 + (x_3 - 1)^2 + (x_4 - 4)^2 \right. \\ \left. + (x_5 - 1)^2\right],$$
$$\text{Minimize } f_2(\mathbf{x}) = x_1^2 + x_2^2 + x_3^2 + x_4^2 + x_5^2 + x_6^2,$$

$$\begin{aligned}
\text{subject to } C_1(\mathbf{x}) &\equiv x_1 + x_2 - 2 \geq 0, \\
C_2(\mathbf{x}) &\equiv 6 - x_1 - x_2 \geq 0, \\
C_3(\mathbf{x}) &\equiv 2 - x_2 + x_1 \geq 0, \\
C_4(\mathbf{x}) &\equiv 2 - x_1 + 3x_2 \geq 0, \quad\quad (7)\\
C_5(\mathbf{x}) &\equiv 4 - (x_3 - 3)^2 - x_4 \geq 0, \\
C_6(\mathbf{x}) &\equiv (x_5 - 3)^2 + x_6 - 4 \geq 0, \\
0 \leq x_1, &x_2, x_6 \leq 10, \quad 1 \leq x_3, x_5 \leq 5, \quad 0 \leq x_4 \leq 6.
\end{aligned}$$

There are six constraints, four of which are linear. Since this is a six-variable problem, it is difficult to show the feasible decision variable space. However, a careful analysis of the constraints and the objective function reveals the constrained Pareto-optimal front, as shown in Fig. 3. The Pareto-optimal region is a concatenation of five regions [8]. Every region lies on some of the constraints. However, for the entire Pareto-optimal region, $x_4^* = x_6^* = 0$. Table 1 shows the other variable values in each of the five regions and the constraints that are active in each region. Since the entire Pareto-optimal region demands an MOEA to maintain its subpopulations at different intersections of constraint boundaries, this may be a difficult problem to solve.

Test Problem SRN Srinivas and Deb [32] borrowed the following function from Chankong and Haimes [3]:

96

K. Deb

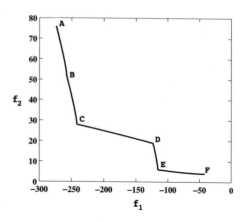

Fig. 3 The constrained
Pareto-optimal front for the
problem OSY. This is a
reprint of Fig. 4 from
Deb et al. (2001)
(© Springer-Verlag Berlin
Heidelberg 2001)

Table 1 Pareto-optimal solutions for the problem OSY. This is a reprint of Table 1 from Deb et al.
(2001) (© Springer-Verlag Berlin Heidelberg 2001)

Region	Optimal values				Active
	x_1^*	x_2^*	x_3^*	x_5^*	constraints
AB	5	1	$(1,\ldots,5)$	5	2, 4, 6
BC	5	1	$(1,\ldots,5)$	1	2, 4, 6
CD	$(4.056,\ldots,5)$	$(x_1^*-2)/3$	1	1	4, 5, 6
DE	0	2	$(1,\ldots,3.732)$	1	1, 3, 6
EF	$(0,\ldots,1)$	$2-x_1^*$	1	1	1, 5, 6

$$\text{Minimize } f_1(\mathbf{x}) = 2 + (x_1-2)^2 + (x_2-1)^2,$$
$$\text{Minimize } f_2(\mathbf{x}) = 9x_1 - (x_2-1)^2,$$
$$\text{subject to } C_1(\mathbf{x}) \equiv x_1^2 + x_2^2 \le 225,$$
$$C_2(\mathbf{x}) \equiv x_1 - 3x_2 + 10 \le 0,$$
$$-20 \le x_1 \le 20,$$
$$-20 \le x_2 \le 20.$$

(8)

Figure 4 shows the feasible decision variable space and the corresponding Pareto-
optimal set [8]. By calculating the derivatives of both objectives and using KKT
optimality conditions, we obtain:

$$\begin{vmatrix} 2(x_1^*-2) & 9 \\ 2(x_2^*-1) & -2(x_2^*-1) \end{vmatrix} = 0.$$

The above equation is satisfied for two cases: (i) $x_1^* = -2.5$ and (ii) $x_2^* = 1$. In
the feasible region, only the first case prevails. Thus, the Pareto-optimal solutions
correspond to $x_1^* = -2.5$ and $x_2^* \in [-14.79, 2.50]$. The feasible objective space,
along with the Pareto-optimal solutions, are shown in Fig. 5. The only difficulty

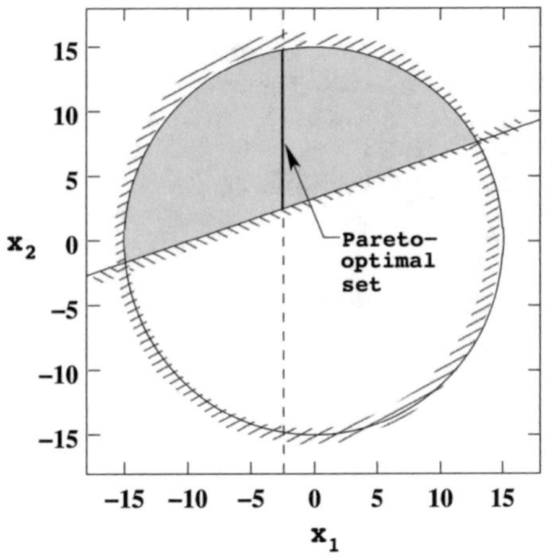

Fig. 4 Constrained decision variable space for the test problem SRN. This is a reprint of Fig. 1 from Deb et al. (2001) (© Springer-Verlag Berlin Heidelberg 2001)

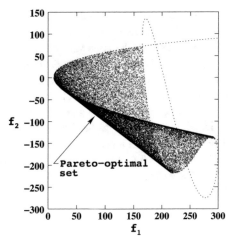

Fig. 5 Feasible objective space for the test problem SRN. This is a reprint of Fig. 2 from Deb et al. (2001) (© Springer-Verlag Berlin Heidelberg 2001)

the constraints introduce in this problem is that they eliminate some parts of the unconstrained Pareto-optimal set (shown by dashed lines).

Test Problem TNK Tanaka [33] suggested the following two-variable problem:

$$
\begin{aligned}
&\text{Minimize } f_1(\mathbf{x}) = x_1, \\
&\text{Minimize } f_2(\mathbf{x}) = x_2, \\
&\text{subject to } C_1(\mathbf{x}) \equiv x_1^2 + x_2^2 - 1 - 0.1 \cos\left(16 \arctan \frac{x_1}{x_2}\right) \geq 0, \\
&\qquad\quad C_2(\mathbf{x}) \equiv (x_1 - 0.5)^2 + (x_2 - 0.5)^2 \leq 0.5, \\
&\qquad\quad 0 \leq x_1 \leq \pi, \\
&\qquad\quad 0 \leq x_2 \leq \pi.
\end{aligned}
\tag{9}
$$

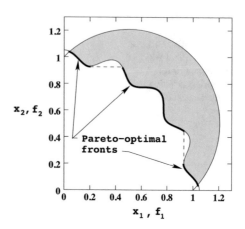

Fig. 6 The feasible decision variable and objective spaces for the TNK problem. This is a reprint of Fig. 3 from Deb et al. (2001) (© Springer-Verlag Berlin Heidelberg 2001)

The feasible decision variable space is shown in Fig. 6. Since $f_1 = x_1$ and $f_2 = x_2$, the feasible objective space is also the same as the feasible decision variable space. The unconstrained decision variable space consists of all solutions in the square $0 \le (x_1, x_2) \le \pi$. Thus, the only unconstrained Pareto-optimal solution is $x_1^* = x_2^* = 0$. However, the inclusion of the first constraint makes this solution infeasible. The constrained Pareto-optimal solutions lie on the boundary of the first constraint. Since the constraint function is periodic and the second constraint function must also be satisfied, not all solutions on the boundary of the first constraint are Pareto-optimal. The disconnected Pareto-optimal set is shown in Fig. 6. Since the Pareto-optimal solutions lie on a nonlinear constraint surface, an optimization algorithm may have difficulty in finding a good spread of solutions across all of the discontinuous Pareto-optimal sets.

4.2 Two-Objective CTP Problems

The above test problems are not tunable for introducing varying degrees of complexity in constrained optimization. As for the tunable unconstrained test problems, we also suggest here a number of test problems where the complexity of the constrained search space can be controlled. The proposed problems are designed to cause two different kinds of tunable difficulties in a multi-objective optimization algorithm, as follows:

– difficulty near the Pareto-optimal front;
– difficulty in the entire search space.

We will discuss both of these in the following subsections.
Difficulty in the Vicinity of the Pareto-Optimal Front In these test problems, constraints do not make any major portion of the search region infeasible, except near the Pareto-optimal front. Constraints are designed in a way so that some portion

of the unconstrained Pareto-optimal region is now infeasible. In this way, the overall Pareto-optimal front will constitute some part of the unconstrained Pareto-optimal region and some part of the constraint boundaries. In the following, we present a generic test problem with J constraints:

$$
\text{CTP1}: \begin{cases} \text{Minimize } f_1(\mathbf{x}_I), \\ \text{Minimize } f_2(\mathbf{x}) = g(\mathbf{x}_{II}) \exp(-f_1(\mathbf{x}_I)/g(\mathbf{x}_{II})), \\ \text{subject to } C_j(\mathbf{x}) \equiv f_2(\mathbf{x}) - a_j \exp[-b_j f_1(\mathbf{x}_I)] \geq 0, \quad j = 1, 2, \ldots, J. \end{cases} \tag{10}
$$

Here, $\mathbf{x} = (\mathbf{x}_I, \mathbf{x}_{II})^T$ and the function $f_1(\mathbf{x}_I)$ and $g(\mathbf{x}_{II})$ can be any multi-variable functions. There are J inequality constraints and the parameters (a_j, b_j) must be chosen in a way so that at least some portion of the unconstrained Pareto-optimal region is infeasible. We now describe a procedure to calculate the (a_j, b_j) parameters for J constraints.

Procedure for Calculating a_j and b_j

Step 1   Set $j = 0$, $a_j = b_j = 1$; also set $\Delta = 1/(J + 1)$ and $\alpha = \Delta$.
Step 2   Calculate $\beta = a_j \exp(-b_j \alpha)$ and

$$
a_{j+1} = (a_j + \beta)/2, \quad b_{j+1} = -\frac{1}{\alpha} \ln(\beta/a_{j+1}).
$$

Increment $\alpha = \alpha + \Delta$ and $j = j + 1$.
Step 3   If $j < J$, go to Step 2. Otherwise, the process is complete.

For two constraints ($J = 2$), the above procedure finds the following parameter values:

$$
a_1 = 0.858, \quad b_1 = 0.541, \quad a_2 = 0.728, \quad b_2 = 0.295.
$$

Figure 7 shows the unconstrained Pareto-optimal region (with a dashed line), and the two constraints. With the presence of both constraints, the figure demonstrates that only about one-third of the original unconstrained Pareto-optimal region is now

Fig. 7 Constrained test problem CTP1 with two constraints

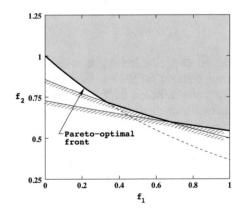

feasible. The other two-thirds part of the constrained Pareto-optimal region comes from the two constraints.

The reason why this problem may cause difficulty to an EMO algorithm is as follows. Since a part of the constraint boundary of each constraint now constitutes the Pareto-optimal region, a spread in the Pareto-optimal solutions requires the decision variables (**x**) to satisfy the inequality constraints with the equality sign. Each constraint is an implicit nonlinear function of the decision variables. Thus, it may be difficult to discover and maintain a number of solutions on a nonlinear constraint boundary. The presence of more numbers of such constraints will demand the algorithm to discover and maintain many such correlations among the decision variables. The complexity of the test problem can be further increased by using a multi-modal function g.

Furthermore, besides finding and maintaining correlated decision variables to fall on several constraint boundaries, there could be other difficulties near the Pareto-optimal front. The constraint functions can be such that the unconstrained Pareto-optimal region is now infeasible and the resulting Pareto-optimal set is a collection of a number of discrete regions. Let us first present such a function mathematically and then describe the difficulties:

$$
\begin{array}{ll}
\text{CTP2--} \\
\text{CTP8 :}
\end{array}
\left\{
\begin{array}{l}
\text{Minimize } f_1(\mathbf{x}) = x_1, \\
\text{Minimize } f_2(\mathbf{x}) = g(\mathbf{x}) \left(1 - \frac{f_1(\mathbf{x})}{g(\mathbf{x})}\right), \\
\text{subject to } C(\mathbf{x}) \equiv \cos(\theta)[f_2(\mathbf{x}) - e] - \sin(\theta) f_1(\mathbf{x}) \geq \\
\qquad\qquad a \left|\sin\left\{b\pi \left[\sin(\theta)(f_2(\mathbf{x}) - e) + \cos(\theta) f_1(\mathbf{x})\right]^c\right\}\right|^d.
\end{array}
\right.
\tag{11}
$$

The decision variable x_1 is restricted in $[0, 1]$ and the bounds of other variables depend on the chosen $g(\mathbf{x})$ function. It is important to note that the problem can be made harder by choosing a multi-variate f_1 function. The constraint $C(\mathbf{x})$ has six parameters (θ, a, b, c, d and e). In fact, the above problem can be used as a constrained test problem generator by tuning these six parameters. We use the above problem to construct different test problems.

First, we use the following parameter values:

$$
\theta = -0.2\pi, \quad a = 0.2, \quad b = 10, \quad c = 1, \quad d = 6, \quad e = 1.
$$

The resulting feasible objective space is shown in Fig. 8. It is clear from this figure that the unconstrained Pareto-optimal region (shown by dashes) is now infeasible. The periodic nature of the constraint boundary makes the Pareto-optimal region discontinuous, having a number of disconnected continuous regions. The task of an optimization algorithm would be to find as many such disconnected regions as possible. The number of such regions can be controlled by increasing the value of the parameter b. It is also clear that with the increase in number of disconnected regions, an algorithm will have difficulty in finding representative solutions in all disconnected regions.

The above problem can be made more difficult by using a small value of d, so that in each disconnected region there exists only one Pareto-optimal solution. Figure 9

Fig. 8 The constrained test problem CTP2

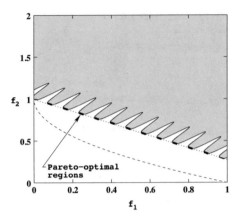

Fig. 9 The constrained test problem CTP3

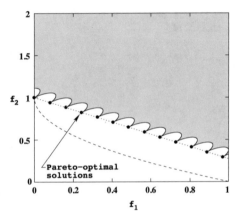

shows the feasible objective space for $d = 0.5$ and $a = 0.1$ (while other parameters are the same as that in the previous test problem). Although most of the search space is feasible, near the Pareto-optimal region the feasible search regions are disconnected, with finally each sub-region leading to a singular feasible Pareto-optimal solution. An algorithm will face difficulty in finding all discrete Pareto-optimal solutions because of the changing nature from a continuous to a discontinuous feasible search space near the Pareto-optimal region.

The problem can be made more difficult by increasing the value of the parameter a, which has an effect of making the transition from the continuous to the discontinuous feasible region far away from the Pareto-optimal region. Since an algorithm now has to travel a long narrow feasible *tunnel* in search of the lone Pareto-optimal solution at the end of the tunnel, this problem will be much more difficult to solve compared to the previous problem. Figure 10 shows one such problem with $a = 0.75$ with the rest of the parameters being the same as those in the previous test problem.

In all of the three above problems, the disconnected regions are equally distributed in the objective space. The discrete Pareto-optimal solutions can be scattered non-

Fig. 10 The constrained test
problem CTP4

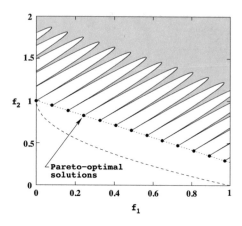

Fig. 11 The constrained test
problem CTP5

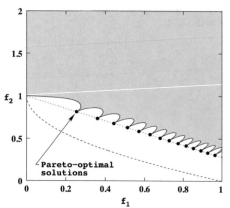

uniformly by using $c \neq 1$. Figure 11 shows the feasible objective space for a problem
with $c = 2$ and with the other parameters the same as those in Fig. 9. Since $c > 1$,
more Pareto-optimal solutions lie towards the right (higher values of f_1). If, however,
$c < 1$ is used, more Pareto-optimal solutions will lie towards the left. For more Pareto-
optimal solutions towards the right, the problem can be made more difficult by using
a large value of c. The difficulty will arise in finding all of the many closely packed
discrete Pareto-optimal solutions.

It is important to mention here that although the above test problems will cause
difficulty in the vicinity of the Pareto-optimal region, an algorithm has to maintain
an adequate diversity well before it comes close to the Pareto-optimal region. If an
algorithm approaches the Pareto-optimal region without much diversity, it may be
too late to create diversity among the population members, as the feasible search
region in the vicinity of the Pareto-optimal region is discontinuous.

Difficulty in the Entire Search Space The above test problems cause difficulty to
an algorithm in the vicinity of the Pareto-optimal region. Difficulties may also come
from the infeasible search region in the entire search space. Fortunately, the same

Fig. 12 The constrained test
problem CTP6

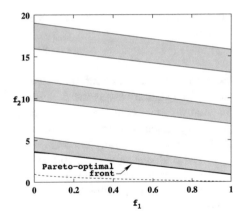

constrained test problem generator can also be used for this purpose. Figure 12 shows
the feasible objective search space for the following parameter values:

$$\theta = 0.1\pi, \quad a = 40, \quad b = 0.5, \quad c = 1, \quad d = 2, \quad e = -2.$$

The objective space has infeasible bands of differing widths towards the Pareto-
optimal region. Since an algorithm has to overcome a number of such infeasible
bands before coming to the island containing the Pareto-optimal front, an MOEA
may face difficulty in solving this problem. The unconstrained Pareto-optimal region
is now not feasible. The entire constrained Pareto-optimal front lies on a part of
the constraint boundary. The difficulty can be increased by widening the infeasible
regions (by using a small value of d).

4.3   Scalable Constrained Test Problem Generator

In the following, we construct a test problem generator having M objective functions, similar in construction to the two-objective test problem generator described
elsewhere [14]. First, we describe the unconstrained version:

$$\left.\begin{array}{l}
\text{Minimize } f_1(\mathbf{x}_1), \\
\text{Minimize } f_2(\mathbf{x}_2), \\
\quad \vdots \\
\text{Minimize } f_{M-1}(\mathbf{x}_{M-1}), \\
\text{Minimize } f_M(\mathbf{x}) = g(\mathbf{x}_M)h\left(f_1(\mathbf{x}_1), f_2(\mathbf{x}_2), \ldots, f_{M-1}(\mathbf{x}_{M-1}), g(\mathbf{x}_M)\right), \\
\text{subject to } \mathbf{x}_i \in \mathbb{R}^{|\mathbf{x}_i|}, \quad \text{for } i = 1, 2, \ldots, M.
\end{array}\right\} \quad (12)$$

Here, the decision variable vector $\mathbf{x}$ is partitioned into M non-overlapping blocks as follows:

$$\mathbf{x} \equiv (\mathbf{x}_1, \mathbf{x}_2, \ldots, \mathbf{x}_{M-1}, \mathbf{x}_M)^T.$$

Each vector $\mathbf{x}_i$ can be of different size. The objective functions $f_1 - f_{M-1}$ can be chosen in a similar way as the function f_1 chosen in the previous subsection. The function g can be similar to the function g described earlier and has the effect of producing difficulty in progressing towards the true Pareto-optimal front. However, the function h is now different and must include all objective function values $f_1 - f_{M-1}$ and g. However, the structure of the function h may be similar to the function h described earlier and has the effect of causing difficulty along the Pareto-optimal front. For example, the following h function will produce a continuous Pareto-optimal region:

$$h(f_1, f_2, \ldots, f_{M-1}, g) = 1 - \left(\frac{\sum_{i=1}^{M-1} f_i}{\beta g} \right)^{\alpha}. \tag{13}$$

The Pareto-optimal front will be convex for $\alpha < 1$. The parameter β is a normalization parameter. In order to create a problem with a discontinuous set of Pareto-optimal fronts, a periodic h function as used in the previous subsection can also be developed here. Since the first $(M-1)$ objectives are functions of the non-overlapping set of decision variables, the Pareto-optimal solutions correspond to the values of $\mathbf{x}_M^*$ for which g is minimum and for all permissible values of the first $(M-1)$ set of variables, which satisfy $f_M = g(\mathbf{x}_M^*)h(f_1, \ldots, f_{M-1}, g(\mathbf{x}_M^*))$ (the Pareto-optimal surface) and are non-dominated by each other.

In order to construct more difficult test problems, the decision variable vector $\mathbf{x}$ can be mapped into a different variable vector $\mathbf{y}$, as suggested above in equation (3).

Next, we suggest a constrained version. Using a M-dimensional transformation (rotational R and translational $\mathbf{e}$) of M-dimensional objective vectors, we compute:

$$\mathbf{f}' = R^{-1}(\mathbf{f} - \mathbf{e}).$$

The matrix R will involve $(M-1)$ rotation angles. Thereafter, the following constraint is added:

$$C(\mathbf{x}) \equiv f_M'(\mathbf{x}) - \sum_{j=1}^{M-1} a_j \left| \sin \left(b_j \pi f'(\mathbf{x})_j^{c_j} \right) \right|^{d_j} \geq 0. \tag{14}$$

Here, a_j, b_j, c_j, d_j and θ_j are all parameters that must be set to get a desired effect. As before, a combination of more than one such constraints can also be used.

4.4   Scalable Constrained DTLZ Problems Using Constraint Surface Concept

Scalable DTLZ test problems were suggested for testing the performance of EMO algorithms beyond two objectives. The first seven test problems were problems with box constraints alone, however, DTLZ8 and DTLZ9 problems involved constraints.

First, we describe the constraint surface concept [14]. The constraint surface approach begins by a predefining the overall search space. Here, a simple geometry such as a rectangular hyper-box is assumed. Each objective function value is restricted to lie within a predefined lower and a upper bound. The resulting problem is as follows:

$$\left.\begin{array}{l} \text{Minimize } f_1(\mathbf{x}), \\ \text{Minimize } f_2(\mathbf{x}), \\ \quad\vdots \qquad\quad \vdots \\ \text{Minimize } f_M(\mathbf{x}), \\ \text{Subject to } f_i^{(L)} \le f_i(\mathbf{x}) \le f_i^{(U)} \quad \text{for } i = 1, 2, \dots, M. \end{array}\right\} \qquad (15)$$

It is intuitive that the Pareto-optimal set of the above problem has only one solution (the solution with the lower bound of each objective $(f_1^{(L)}, f_2^{(L)}, \dots, f_1^{(L)})^T$. Figure 13 shows this problem for three objectives (with $f_i^{(L)} = 0$ and $f_i^{(U)} = 1$) and the resulting singleton Pareto-optimal solution $\mathbf{f} = (0, 0, 0)^T$) is also marked.

The problem is now made more interesting by adding a number of constraints (linear or non-linear):

$$g_j(f_1, f_2, \dots, f_M) \ge 0, \quad \text{for } j = 1, 2, \dots, J. \qquad (16)$$

This is done to chop off portions of the original rectangular region systematically. Figure 14 shows the resulting feasible region after adding the following two linear constraints:

$$g_1 \equiv f_1 + f_3 - 0.5 \ge 0,$$
$$g_2 \equiv f_1 + f_2 + f_3 - 0.8 \ge 0.$$

What remains is the feasible search space. The objective of the above problem is to find the non-dominated portion of the boundary of the feasible search space. Figure 14 also marks the Pareto-optimal surface of the above problem. For simplicity and easier comprehension, each constraint involving at most two objectives (similar to the first constraint above) can be used.

Test Problem DTLZ8 The constraint surface approach is used to construct the following test problem:

Fig. 13 Entire cube is the
search space. The origin is
the sole Pareto-optimal
solution

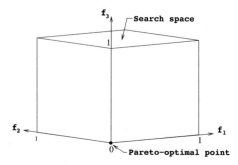

Fig. 14 Two constraints are
added to eliminate a portion
of the cube, thereby resulting
in a more interesting
Pareto-optimal front

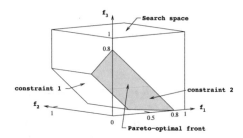

$$\left.\begin{array}{l} \text{Minimize } f_j(\mathbf{x}) = \frac{1}{\lfloor \frac{n}{M} \rfloor} \sum_{i=\lfloor (j-1)\frac{n}{M} \rfloor}^{\lfloor j\frac{n}{M} \rfloor} x_i, \quad j = 1, 2, \ldots, M, \\ \text{Subject to } g_j(\mathbf{x}) = f_M(\mathbf{x}) + 4f_j(\mathbf{x}) - 1 \geq 0, \quad \text{for } j = 1, 2, \ldots, (M-1), \\ \qquad g_M(\mathbf{x}) = 2f_M(\mathbf{x}) + \min_{\substack{i,j=1 \\ i \neq j}}^{M-1} \left[f_i(\mathbf{x}) + f_j(\mathbf{x}) \right] - 1 \geq 0, \\ \qquad 0 \leq x_i \leq 1, \quad \text{for } i = 1, 2, \ldots, n. \end{array}\right\}$$

$$(17)$$

Here, the number of variables is considered to be larger than the number of objectives, or $n > M$. We suggest $n = 10M$. In this problem, there are a total of M constraints. The Pareto-optimal front is a combination of a straight line and a hyper-plane. The straight line is the intersection of the first $(M-1)$ constraints (with $f_1 = f_2 = \cdots = f_{M-1}$) and the hyper-plane is represented by the constraint g_M. MOEAs may find difficulty in finding solutions in both the regions in this problem and also in maintaining a good distribution of solutions on the hyper-plane. Figures 15 and 16 show NSGA-II ands SPEA2 populations after 500 generations. The Pareto-optimal region (a straight line and a triangular plane) is also marked in the plots. Although some solutions on the true Pareto-optimal front are found, there exist many other non-dominated solutions in the final population. These redundant solutions lie on the adjoining surfaces to the Pareto-optimal front. Their presence in the final non-dominated set is difficult to eradicate in real-parameter MOEAs, a matter which we discuss in the next subsection.

Test Problem DTLZ9 This test problem is also created using the constraint surface approach:

Fig. 15 The NSGA-II
population of non-dominated
solutions on test problem
DTLZ8

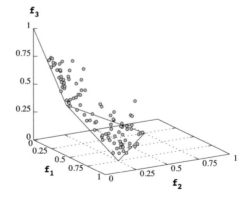

Fig. 16 The SPEA2
population of non-dominated
solutions on test problem
DTLZ8

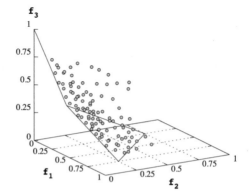

$$\left.\begin{array}{l} \text{Minimize } f_j(\mathbf{x}) = \sum_{i=\lfloor(j-1)\frac{n}{M}\rfloor}^{\lfloor j\frac{n}{M}\rfloor} x_i^{0.1}, \quad j = 1, 2, \dots, M, \\ \text{Subject to } g_j(\mathbf{x}) = f_M^2(\mathbf{x}) + f_j^2(\mathbf{x}) - 1 \geq 0, \quad \text{for } j = 1, 2, \dots, (M-1), \\ \quad 0 \leq x_i \leq 1, \quad \text{for } i = 1, 2, \dots, n. \end{array}\right\}$$

$$(18)$$

Here too, the number of variables is considered to be larger than the number of
objectives. For this problem, we suggest $n = 10M$. The Pareto-optimal front is a
curve with $f_1 = f_2 = \dots = f_{M-1}$, similar to that in DTLZ5. However, the density
of solutions gets thinner towards the Pareto-optimal region. The Pareto-optimal curve
lies on the intersection of all $(M-1)$ constraints. This feature of this problem may
cause MOEAs difficulty in solving this problem. However, the symmetry of the
Pareto-optimal curve in terms of $(M-1)$ objectives allows an easier way to illustrate
the obtained solutions. A two-dimensional plot of the Pareto-optimal front with f_M
and any other objective function should represent a circular arc of radius one. A
plot with any two objective functions except f_M should show a 45° straight line.
Figures 17 and 18 show NSGA-II and SPEA2 populations after 500 generations on
a f_3-f_1 plot of the 30-variable, three-objective DTLZ9 problem. The Pareto-optimal
circle is also shown in the plots. It is clear that both algorithms could not cover the

Fig. 17 The NSGA-II
population on test problem
DTLZ9

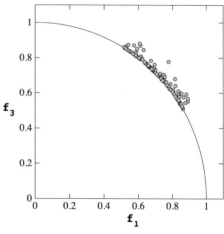

Fig. 18 The SPEA2
population on test problem
DTLZ9

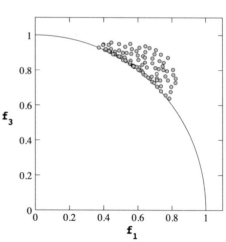

entire range of the circle and there exist many non-dominated solutions away from
the Pareto-optimal front.

4.5 Constrained Test Problems for Many-Objective
Optimization

Recently proposed constrained test problems for many-objective optimization [22]
are scalable from two to as many objectives as desired. We describe them here.

Constrained Problems of Type-1 In Type-1 constrained problems, the original Pareto-optimal front still remain optimal, but there is an infeasible barrier in
approaching the Pareto-optimal front. This is achieved by adding a constraint to the

original problem. The barrier provides infeasible regions in the objective space that an algorithm must learn to overcome, thereby providing a difficulty in converging to the true Pareto-optimal front. DTLZ1 and DTLZ3 problems of the DTLZ family of unconstrained test problems [14] are modified according to this principle.

For the type 1 constrained DTLZ1 (or C1-DTLZ1), only a part of objective space that is close to Pareto-front is made feasible, as shown in Fig. 19. The objective functions are kept the same as they were in the original DTLZ1 problem, while the following constraint is now added:

$$c(\mathbf{x}) = 1 - \frac{f_M(\mathbf{x})}{0.6} - \sum_{i=1}^{M-1} \frac{f_i(\mathbf{x})}{0.5} \geq 0. \tag{19}$$

The feasible region and the Pareto-optimal front are shown for a two-objective C1-DTLZ1 problem in Fig. 19. In all simulations, we use $k = 5$ variables for the original g-function [14], thereby making a total of $(M + 4)$ variables to the M-objective C1-DTLZ1 problem.

In the case of C1-DTLZ3 problem, a band of infeasible space is introduced adjacent to the Pareto-optimal front, as shown in Fig. 20. Again, the objective functions are kept the same as in original DTLZ3 problem [14], while the following constraint is added:

$$c(\mathbf{x}) = \left(\sum_{i=1}^{M} f_i(\mathbf{x})^2 - 16 \right) \left(\sum_{i=1}^{M} f_i(\mathbf{x})^2 - r^2 \right) \geq 0. \tag{20}$$

where, $r = \{9, 12.5, 12.5, 15, 15\}$ is the radius of the hyper-sphere for $M = \{3, 5, 8, 10, 15\}$. For C1-DTLZ3, we use $k = 10$, so that total number of variables are $(M + 9)$ in a M-objective problem.

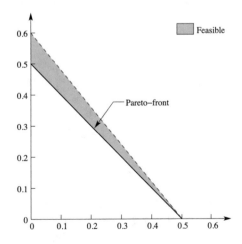

Fig. 19 Two objective version of C1-DTLZ1 problem

Fig. 20 Two objective
version of C1-DTLZ3
problem

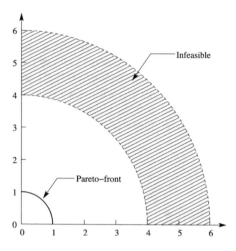

4.6 Constrained Problems of Type-2

While type 1 constrained problems introduce difficulties in arriving at the entire
Pareto-optimal front, type-2 constrained problems are designed to introduce infeasibility to a part of the Pareto-optimal front. Such problems will test an algorithm's
ability to deal with disconnected Pareto-optimal fronts. To accomplish this, DTLZ2
[14] and the convex DTLZ2 problems [12] are modified.

In C2-DTLZ2 problem, only the region of objective space that lies inside each of
the $M + 1$ hyper-spheres of radius r is made feasible. Of $(M + 1)$ hyper-spheres,
M are placed at the corners of unit hyper-plane and the $(M + 1)$-th is placed at the
intersection of the equally-angled line with objective axes and the original Pareto-
optimal front. This way, the Pareto-optimal front is disconnected, as shown in Fig. 21.
Objective functions are calculated in the same way as in the original DTLZ2 problem,
except that a constraint is now introduced:

$$
c(\mathbf{x}) = -\min\left\{ \min_{i=1}^{M}\left[(f_i(\mathbf{x}) - 1)^2 + \sum_{j=1,j\neq i}^{M} f_j^2 - r^2 \right], \right.
$$
$$
\left. \left[\sum_{i=1}^{M} (f_i(\mathbf{x}) - 1/\sqrt{M})^2 - r^2 \right] \right\} \geq 0,
$$

where $r = 0.4$, for $M = 3$ and 0.5, otherwise. For an M-objective C2-DTLZ2 problem, $k = 10$ is used, thereby having a total of $(M + 9)$ variables.

For the convex C2-DTLZ2 described in [12], we construct a different feasible region. The region in the objective space lying inside a hyper-cylinder with
$(1, 1, \ldots, 1)^T$ as the axis and radius r is kept infeasible, thereby creating an infeasible hole through the objective space. This also produces a hole on the Pareto-optimal

Fig. 21 Two-objective
version of C2-DTLZ2
problem

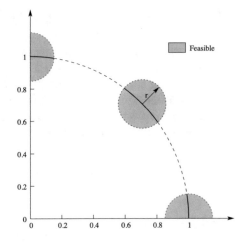

Fig. 22 Two-objective
version of convex
C2-DTLZ2 problem

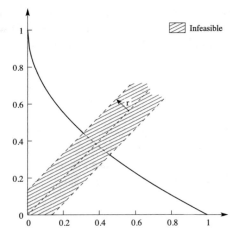

front, as demonstrated for a two-objective version of convex C2-DTLZ2 problem
in Fig. 22. The objective functions are kept the same as before, while the following
constraint is added:

$$c(\mathbf{x}) = \sum_{i=1}^{M} (f_i(\mathbf{x}) - \lambda)^2 - r^2 \geq 0, \tag{21}$$

where $\lambda = \frac{1}{M} \sum_{i=1}^{M} f_i(\mathbf{x})$ and the radius $r = \{0.225, 0.225, 0.26, 0.26, 0.27\}$ for
$M = \{3, 5, 8, 10, 15\}$. Total number of variables for this problem are $(M + 9)$.

4.7 Constrained Problems of Type-3

Type-3 problems involve multiple constraints and the entire Pareto-optimal front of the unconstrained problem need not be optimal any more, rather portions of the added constraint surfaces constitute the Pareto-optimal front. DTLZ1 and DTLZ4 problems are modified for this purpose by adding M different constraints. In the case of C3-DTLZ1 problem, objective functions are same as in the original formulation [14], however, following M linear constraints are added:

$$c_j(\mathbf{x}) = \sum_{i=1,i\neq j}^{M} f_j(\mathbf{x}) + \frac{f_i(\mathbf{x})}{0.5} - 1 \geq 0, \quad \forall j = 1, 2, \ldots, M. \tag{22}$$

For C3-DTLZ1 problem, $k = 5$ is used in the original g-function, thereby making a total of $(M + 4)$ variables. Figure 23 shows constraints and feasible region for the two-objective C3-DTLZ1 problem. Notice how the unconstrained Pareto-optimal front is now infeasible by the presence of two constraints.

Similarly, DTLZ4 problem is modified by adding M quadratic constraints of the type:

$$c_j(\mathbf{x}) = \frac{f_j^2}{4} + \sum_{i=1,i\neq j}^{M} f_i(\mathbf{x})^2 - 1 \geq 0, \quad \forall j = 1, 2, \ldots, M. \tag{23}$$

An additional difficulty posed by DTLZ4 is that it introduces bias for creating solutions in certain parts of the objective space. For this problem, we suggest $n = M + 4$ variables. Figure 24 shows the respective constraints and resulting Pareto-optimal front for the C3-DTLZ4 problem.

Fig. 23 Two objective version of C3-DTLZ1 problem

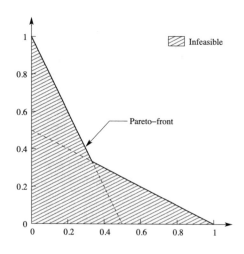

Fig. 24 Two objective
version of C3-DTLZ4
problem

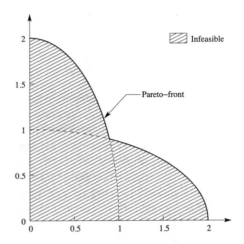

4.8 Other Constrained Problems

In 2009, a set of constrained test problems were suggested for two and three objectives
[36]. The problems can be classified into two classes: (i) one that make some parts of
the original unconstrained Pareto-optimal fronts feasible and (ii) new partial fronts
which lie on one of the constraint boundaries become Pareto-optimal. Moreover,
these problems involve subfunctions in the objective function that forces the optimal
variable combinations to fall on certain non-linear surfaces. This requirement makes
these problem difficult even for their unconstrained versions. Although the effort is
in the right direction, there are a number of issues which we discuss next.

If a problem has an infeasible region from the interior search space all the way to
the Pareto-optimal front in a regular pattern, an algorithm gets a number of iterations
to learn the pattern and eventually is capable of converging to the feasible Pareto-
optimal front. An algorithm must have to be tested for its ability to go through
different types and shapes of infeasibility from interior to the optimal portion of the
search space, for it to be effective in arbitrary practical problems.

As mentioned above, the above test problems make the convergence to the Pareto-
optimal front difficult due to the restricted non-linear relationships that certain variables must satisfy. When constraints are introduced in such problems and a constrained EMO algorithm is unable to solve the problem well, it is difficult to judge
the efficacy of the constrained handling part of the algorithm. Thus, to make a better evaluation of the constraint handling part of an EMO algorithm, test problems
that test an algorithm's ability to negotiate constraints must be designed, keeping
other aspects as benign as possible. Moreover, these test problems are not scalable to
higher dimensions. Efforts must be spent to design scalable constrained multi- and
many-objective test problems.

5 Future Studies in Constrained EMO

In addition to the above proposed methods, in this section, we suggest a number of future directions for research in constrained EMO studies.

1. **Providing More Preference to Infeasible solutions**. Like in single-objective problems, infeasible solutions in a multi-objective problem, if allowed to compete with feasible solutions based on their objective values, can be evaluated to be better. This requires a careful comparison of feasible with infeasible solutions. Author's constrained tournament selection avoids such a comparison completely. However, certain infeasible solutions lying close to the Pareto-optimal front may lead the search to converge quicker to the Pareto-optimal solutions than feasible solutions that may be far away from the Pareto-optimal front. Thus, there is a need to modify the constrained tournament selection operator with a distance based strategy from the current non-dominated solutions in situations where providing more preference to infeasible solutions may lead to a faster search.

2. **Constrained violation is difficult to compute**. The constrained tournament selection operator uses an aggregate constraint violation value to indicate the extent of violation all constraints. However, other constraint violation estimation methods that will rank solutions with fewer constraint infeasibility better than solutions with larger number of constraint infeasibility can be used. Such a method will emphasize and maintain solutions around each constraint boundary. For multi-objective optimization problems, such a method may work better due to the need for finding a set of Pareto-optimal solutions each of which is most likely to lie on a single constraint boundary.

3. **Dimensionality issues with constraints**. Large-scale multi-objective optimization studies usually involve an increasing number of (i) variables and (ii) objectives. But, an increase in the number of constraints may also cause a constrained EMO algorithm difficulty. Thus, in addition to testing an EMO algorithm's scalability with number of variables and objectives, they should also be tested for their efficacy in handling an increasing number of constraints. Unlike in single-objective constrained problems, constraints in a multi-objective problem can provide two types of difficulties: (i) constraints that cause hindrance for convergence towards the Pareto-optimal front and force an EMO algorithm to get stuck at constraint boundaries which are far away from the true Pareto-optimal front, and (ii) constraints that directly make the unconstrained Pareto-optimal front infeasible, thereby making the constrained Pareto-optimal front to lie on constraint boundaries. An EMO algorithm's performance thus must be tested on both types of constraints and their increasing dimensionality in a systematic manner.

4. **Constrained test problems**. The above-mentioned testing of an EMO algorithm requires carefully designed test problems providing both types of constraints and a systematic update to test an algorithm's performance on an increasing number of constraints. Current test problems do not have these aspects and there lies a need for developing such test problems.

5. **Single-objective ideas to multi-objective optimization**. Constraint handling has received a lot of attention in single-objective optimization studies [5]. Since these techniques deal with handling constraints directly without the involvement of objectives per se, some of these techniques may be easily introduced in an EMO algorithm. Of the suggested ideas, the ones which do not require any additional parameter tuning would be the most appropriate ideas to try first. A recently proposed single-objective constrained handling strategy has outperformed a few other state-of-the-art strategies in solving a number of commonly-used test problems by one and two to three orders of magnitude in terms of overall required solution evaluations [10]. In this approach, an additional helper objective (minimization of overall constraint violation) is introduced to obtain an appropriate penalty parameter and a penalty function approach is used to find an improved solution. The method can be extended for multi-objective optimization problems. Its efficacy must be investigated over existing multi-objective EMO methods on standard test problems.

6. **Multi-objective approaches to apply to many-objective optimization**. The recent focus in EMO studies is on solving many-objective optimization problems involving four or more conflicting objectives. Handling constraints in four or more objective problems may require different constrained handling strategies than handling constraints in two and three-objective problems. With appropriate test problems developed and briefly discussed in Sect. 4.5, the constrained EMO algorithms must then be tested for their efficacy in many-objective constrained problems.

7. **Constrained EMO algorithms for multi-objective bilevel problems solving**. Bilevel problems involve an upper and a lower level optimization problem. An upper level solution is considered feasible only if it is associated with the optimal solution of the lower level problem. This makes bilevel optimization problems *nested* and difficult to solve. In the presence of multiple conflicting objectives in each level, the problems get even harder to solve. However, due to the direct relationship of bilevel optimization with constrained optimization, efficient constrained EMO algorithms must be developed to solve multi-objective bilevel problems.

6 Conclusions

Most EMO algorithms have been developed to solve unconstrained multi-objective optimization problems or multi-objective optimization problems involving box constraints only. However, when constraints are present, some of the proposed EMO algorithms may not perform well and new methodologies are required. In this paper, we have presented generic constrained multi-objective optimization algorithms which are commonly used. Optimization algorithms must be tested on carefully designed test problems which systematically evaluate the efficiency of an algorithm in handling different complexities associated with a practical problem. We have discussed the studies associated with development of constrained test problems for multi and many-objective optimization.

Although this chapter has accounted for existing algorithms and test problems for constrained optimization, multi-objective constrained optimization has not been adequately paid attention by the EMO community. In this chapter, we have suggested a few future directions for research and further development in this area.

At the end, we would like to end this chapter by highlighting the fact that any optimization algorithm is not complete and not pragmatic without an efficient way of handling constraints. Although an algorithm can be developed first to handle box constraints alone, it must be accompanied with an adequate constrained handling procedure if the proposed algorithm is to be accepted in practice. Hopefully, this chapter motivates EMO researchers in this direction and emphasizes the need for paying more attention to constrained handling strategies in an EMO algorithm.

References

1. Bhatia, D., Aggarwal, S.: Optimality and duality for multiobjective nonsmooth programming. Eur. J. Oper. Res. **57**(3), 360–367 (1992)
2. Binh, T.T., Korn, U.: MOBES: A multiobjective evolution strategy for constrained optimization problems. In: The Third International Conference on Genetic Algorithms (Mendel 97), pp. 176–182 (1997)
3. Chankong, V., Haimes, Y.Y.: Multiobjective Decision Making Theory and Methodology. North-Holland, New York (1983)
4. Da Cunha, N.O., Polak, E.: Constrained minimization under vector-evaluated criteria in finite dimensional spaces. J. Math. Anal. Appl. **19**(1), 103–124 (1967)
5. Datta, R., Deb, K. (eds.): Evolutionary Constrained Optimization. Infosys Science Foundation Series, Springer (2015)
6. Deb, K.: Optimization for Engineering Design: Algorithms and Examples. Prentice-Hall, New Delhi (1995)
7. Deb, K.: Evolutionary algorithms for multi-criterion optimization in engineering design. In: Miettinen, K., Neittaanmäki, P., Mäkelä, M.M., Périaux, J. (eds.) Evolutionary Algorithms in Engineering and Computer Science, pp. 135–161. Wiley, Chichester (1999)
8. Deb, K.: Multi-objective Optimization Using Evolutionary Algorithms. Wiley, Chichester (2001)
9. Deb, K., Agrawal, S., Pratap, A., Meyarivan, T.: A fast and elitist multi-objective genetic algorithm: NSGA-II. IEEE Trans. Evol. Comput. **6**(2), 182–197 (2002)
10. Deb, K., Datta, R.: A fast and accurate solution of constrained optimization problems using a hybrid bi-objective and penalty function approach. In: Proceedings of the IEEE World Congress on Computational Intelligence (WCCI-2010), pp. 165–172 (2010)
11. Deb, K., Goldberg, D.E.: An investigation of niche and species formation in genetic function optimization. In: Proceedings of the Third International Conference on Genetic Algorithms, pp. 42–50 (1989)
12. Deb, K., Jain, H.: An improved NSGA-II procedure for many-objective optimization Part I: Problems with box constraints. Technical Report 2012009, Indian Institute of Technology Kanpur (2012)
13. Deb, K., Jain, H.: An evolutionary many-objective optimization algorithm using reference-point based non-dominated sorting approach, Part I: solving problems with box constraints. IEEE Trans. Evol. Comput. **18**(4), 577–601 (2014)

14. Deb, K., Thiele, L., Laumanns, M., Zitzler, E.: Scalable test problems for evolutionary multi-objective optimization. In: Abraham, A., Jain, L., Goldberg, R. (eds.) Evolutionary Multiobjective Optimization, pp. 105–145. Springer, London (2005)
15. Drechsler, R.: Evolutionary Algorithms for VLSI CAD. Kluwer Academic Publishers, Boston (1998)
16. Ehrgott, M.: Multicriteria Optimization. Springer, Berlin (2005)
17. Fonseca, C.M., Fleming, P.J.: Genetic algorithms for multiobjective optimization: formulation, discussion, and generalization. In: Proceedings of the Fifth International Conference on Genetic Algorithms, pp. 416–423. Morgan Kaufmann, San Mateo (1993)
18. Fonseca, C.M., Fleming, P.J.: Multiobjective optimization and multiple constraint handling with evolutionary algorithms-Part I: A unified formulation. IEEE Trans. Syst. Man Cybern. Part A Syst. Hum. 28(1), 26–37 (1998)
19. Homaifar, A., Lai, S.H.-V., Qi, X.: Constrained optimization via genetic algorithms. Simulation 62(4), 242–254 (1994)
20. Horn, J., Nafploitis, N., Goldberg, D.E.: A niched Pareto genetic algorithm for multi-objective optimization. In: Proceedings of the First IEEE Conference on Evolutionary Computation, pp. 82–87 (1994)
21. Huband, S., Barone, L., While, L., Hingston, P.: A scalable multi-objective test problem toolkit. In: Proceedings of the Evolutionary Multi-Criterion Optimization (EMO-2005). Springer, Berlin (2005)
22. Jain, H., Deb, K.: An evolutionary many-objective optimization algorithm using reference-point based non-dominated sorting approach, Part II: Handling constraints and extending to an adaptive approach. IEEE Trans. Evol. Comput. 18(4), 602–622 (2014)
23. Khare, V., Yao, X., Deb, K.: Performance scaling of multi-objective evolutionary algorithms. In: Proceedings of the Second Evolutionary Multi-Criterion Optimization (EMO-03) Conference (LNCS 2632), pp. 376–390 (2003)
24. Knowles, J.D., Corne, D.W.: Approximating the non-dominated front using the Pareto archived evolution strategy. Evol. Comput. J. 8(2), 149–172 (2000)
25. Michalewicz, Z.: Genetic Algorithms + Data Structures = Evolution Programs. Springer, Berlin (1992)
26. Michalewicz, Z., Schoenauer, M.: Evolutionary algorithms for constrained parameter optimization problems. Evol. Comput. J. 4(1), 1–32 (1996)
27. Miettinen, K.: Nonlinear Multiobjective Optimization. Kluwer, Boston (1999)
28. Osyczka, A., Kundu, S.: A new method to solve generalized multicriteria optimization problems using the simple genetic algorithm. Struct. Optim. 10(2), 94–99 (1995)
29. Ray, T., Tai, K., Seow, K.C.: An evolutionary algorithm for multiobjective optimization. Eng. Optim. 33(3), 399–424 (2001)
30. Reklaitis, G.V., Ravindran, A., Ragsdell, K.M.: Engineering Optimization Methods and Applications. Wiley, New York (1983)
31. Shukla, P., Deb, K.: On finding multiple Pareto-optimal solutions using classical and evolutionary generating methods. Eur. J. Oper. Res. (EJOR) 181(3), 1630–1652 (2007)
32. Srinivas, N., Deb, K.: Multi-objective function optimization using non-dominated sorting genetic algorithms. Evol. Comput. J. 2(3), 221–248 (1994)
33. Tanaka, M.: GA-based decision support system for multi-criteria optimization. In: Proceedings of the International Conference on Systems, Man and Cybernetics vol. 2, pp. 1556–1561 (1995)
34. Van Veldhuizen, D.: Multiobjective Evolutionary Algorithms: Classifications, Analyses, and New Innovations. Ph.D. thesis, Dayton, OH: Air Force Institute of Technology (1999). Technical Report No. AFIT/DS/ENG/99-01
35. Zhang, Q., Li, H.: MOEA/D: a multiobjective evolutionary algorithm based on decomposition. IEEE Trans. Evol. Comput. 11(6), 712–731 (2007)
36. Zhang, Q., Zhou, A., Zhao, S.Z., Suganthan, P.N., Liu, W., Tiwari, S.: Multiobjective optimization test instances for the CEC-2009 special session and competition. Nanyang Technological University, Technical report, Singapore (2008)

37. Zitzler, E., Deb, K., Thiele, L.: Comparison of multiobjective evolutionary algorithms: empirical results. Evol. Comput. J. **8**(2), 125–148 (2000)
38. Zitzler, E., Thiele, L.: Multiobjective evolutionary algorithms: a comparative case study and the strength Pareto approach. IEEE Trans. Evol. Comput. **3**(4), 257–271 (1999)

Genetic Programming for Classification and Feature Selection

Kaustuv Nag and Nikhil R. Pal

Abstract Our objective is to provide a comprehensive introduction to Genetic Programming (GP) primarily keeping in view the problem of classifier design along with feature selection. We begin with a brief account of how genetic programming has emerged as a major computational intelligence technique. Then, we analyse classification and feature selection problems in brief. We provide a naive model of GP-based binary classification strategy with illustrative examples. We then discuss a few existing methodologies in brief and three somewhat related but different strategies with reasonable details. Before concluding, we make a few important remarks related to GP when it is used for classification and feature selection. In this context, we show some experimental results with a recent GP-based approach.

Keywords Classification · Feature selection · Genetic programming

1 Introduction

1.1 The Emergence of Genetic Programming

Computational Intelligence (*CI*) deals with biologically and linguistically inspired computing paradigms. Evolutionary computation (EC) is one of the major components of CI. *Evolutionary algorithms* (EAs), which are concerned with EC, exploit *Darwinian principles* to find solutions to a problem. These are, indeed, trial and error based optimization schemes that use metaheuristics. Moreover, EAs use a population consisting of a set of candidate solutions instead of iterating over a single solution in the search space. Usually, the following four techniques are categorized as

K. Nag (✉)
Department of IEE, Jadavpur University, Kolkata, India
e-mail: kaustuv.nag@gmail.com

N. R. Pal
ECS Unit, Indian Statistical Institute, Calcutta, India
e-mail: nrpal59@gmail.com

© Springer International Publishing AG, part of Springer Nature 2019
J. C. Bansal et al. (eds.), *Evolutionary and Swarm Intelligence Algorithms*, Studies in Computational Intelligence 779,
https://doi.org/10.1007/978-3-319-91341-4_7

EC: (i) *evolutionary programming* (EP), (ii) *evolutionary strategy* (ES), (iii) *genetic algorithm* (GA), and (iv) *genetic programming* (GP).

EC usually initializes a population with a set of randomly generated candidate solutions. However, if domain knowledge is available, it can be used to generate the initial population. Then, the population is *evolved*. This evolutionary process incorporates *natural selection* and other *evolutionary* operators. From an algorithmic point of view, it is a guided random search that uses parallel processing to achieve the desired solutions. Note that, the *natural selection* must be incorporated in an EA, otherwise the approach cannot be categorized as an EC technique. For example, though several metaheuristic algorithms, such as, particle swarm optimization (PSO) [12] and ant colony optimization (ACO) [6, 9] are *nature inspired algorithms* (NIAs), they are not EAs. Note that, sometimes they are still loosely referred to as EC techniques.

In 1948, in a technical report [1], titled "Intelligent Machinery", written for National Physics Laboratory, Alan M. Turing wrote, "There is the genetical or evolutionary search by which a combination of genes is looked for, the criterion being survival value. The remarkable success of this search confirms to some extent the idea that intellectual activity consists mainly of various kinds of search." According to the best of our knowledge, this is the first technical article, where the concept of evolutionary computation is found. However, it took few more decades to develop the following three distinct interpretations of this philosophy: (i) EP, (ii) ES, and (iii) GA. For the next one and half decades, these three areas grew separately. Later, in the early nineties, they were unified as a subfield of the same technology, namely EC. Each of EP, ES and GA is an algorithm for finding solutions to an optimization problem - it finds a parameter vector that optimizes an objective function. Unlike these three branches, GP finds a program to solve a problem. The concept of modern *tree-based* GP was proposed by Cramer in 1985 [7]. Later, Koza, a student of Cramer, popularized it with his many eminent works [15–18]. A large number of GP-based inventions have been made after 2000, i.e., after the emergence of sufficiently well performing hardware.

1.2 Genetic Programming: The Special Encoding Scheme

GP finds computer programs to perform a given task. GP consists of a set of instructions and a fitness function to evaluate the performance of a candidate computer program. It can be considered a special case of GA, where each solution is a computer program. Traditionally for GP computer programs are represented in the memory as tree structures [7]. The internal nodes of the trees must be from a set of predefined functions (operators), $\mathcal{F}$. Moreover, every leaf node of a tree must be from a set of predefined terminals (operands), $\mathcal{T}$. The subtrees of a function node $f \in \mathcal{F}$ are the arguments to f. Note that, a very important property of tree-based GP is that $\mathcal{F}$ and $\mathcal{T}$ need to satisfy both the *closure* property and the *sufficiency* property [15]. To satisfy the closure property, $\mathcal{F}$ needs to be well defined and closed for any combination of probable arguments that it may encounter [15]. Moreover, $\mathcal{F}$ and

$\mathcal{T}$ need to be able to encode any possible valid solution of the problem to satisfy the sufficiency property. Thus, Lisp or any other functional programming language that naturally embody tree structures, can be used to represent a candidate solution in GP. Use of non-tree representations to encode solutions in GP, is comparatively less popular. An example of this is linear genetic programming, which is suitable for more traditional imperative languages. In this chapter, however, we concentrate only on *tree-based* GP. The most frequently used representations of tree-based GPs are decision trees, classification rules, and discriminant functions. Here, we confine our discussion primarily to discriminant function based GPs.

1.3 Classification and Feature Selection

In machine learning and data mining, *classification* is an important and frequently encountered problem. Besides, it is possible to restate a wide range of real world problems as classification problems. Examples of such real world problems include medical diagnosis, text categorization, and software quality assurance. In a classification task, we have a set $\mathcal{X}$ with N_p training data points, where each element of the set is denoted by a pair $(\mathbf{x}_i, y_i)$, $i = 1, 2, \ldots, N_p$. Here, $\mathbf{x}_i = (x_{1i}, x_{2i}, \ldots, x_{ni})^T \in \mathbb{R}^n$ and $y_i \in Y$, where Y is the set of all class labels. Now, based on the information hidden in $\mathcal{X}$, we need to model a set of criteria in a solver, called classifier, such that, given a point $\mathbf{x} \in \mathbb{R}^n$, the classifier would predict an appropriate class label $y \in Y$.

To encode the set of criteria, a classifier relies on a set of features. Thus, *feature selection* (FS) becomes an important part in classification. FS can be defined as a process of selecting a subset of relevant features that can solve the problem at hand satisfactorily. FS is primarily used for the following three reasons: (i) to simplify models to enhance their interpretability, (ii) to attain shorter training time, and (iii) to enhance the generalization capability of the classifier by reducing its degrees of freedom. Note that, a genetic program may not (mostly will not) use all features of a given data set. Hence, a GP-based system performs FS implicitly, at least to some extent even if it is not specially designed for FS. Moreover, a discriminant function based genetic program also implicitly performs *feature extraction* (FE) from an initial set of measured features. The derived features are expected to be less-redundant, informative, and should facilitate subsequent learning and enhance generalization. Sometimes, they may lead to better human interpretation.

For a given classification task, there may be at least four types of features: (i) essential, (ii) bad, (iii) redundant, and (iv) indifferent [5]. The objective of a FS scheme should be to (i) select the essential features, (ii) reject the bad features, (iii) judiciously select some of the redundant features, and (iv) reject the indifferent features. Let us consider a small example to illustrate these four types of features [5]. Consider a data set on humans with five features: (i) sex, (ii) eye color, (iii) height, (iv) weight, and (v) number of legs. Suppose the given classification task has the following four classes:

(i) male AND (heavy weight OR long height),
(ii) male AND (low weight OR short height),
(iii) female AND (heavy weight OR long height), and
(iv) female AND (low weight OR short height).

In this particular case, (i) the feature *sex* is essential, (ii) the feature *eye color* is bad as it may confuse the learning, (iii) the feature *height* is redundant with feature *weight* as a long person is usually heavy. Thus, *weight* and *height* constitute a redundant set for the given classification task because *usually* any one of the two will be enough for the classification task. Note that, we have emphasized on the word *usually* because height and weight are statistically strongly correlated, but there could be a heavy person with a short height. (iv) Finally, *number of legs* is an indifferent feature as under normal circumstances, it is going to be two for every individual. Note that, keeping some level of redundancy in the selected set of features may sometimes be helpful. Therefore, in the given context, one may argue that to account for some measurement error, it may not be a very bad idea to use both *height* and *weight* to design a classifier because that would be more robust than a classifier designed using only one of these two features. Again, if we want to employ FE in the above described scenario, it would be good to construct a feature *height* '*OR*' *weight*. Note that, this '*OR*' operator is not exactly the Boolean OR operator because the attribute heavy, long etc. are not Boolean, but fuzzy concepts, and hence, such a combined feature has to be designed judiciously. The beauty of a GP-based system is that, during the evolution, it may compute the intuitive '*OR*' operator using the members of $\mathcal{F}$.

1.4 Genetic Programming for Classification and Feature Selection: A Simple Example

Consider the tiny, artificial, two dimensional, binary classification data set shown in Fig. 1. It consists of uniformly distributed points inside two circles: $C_1 : (x_1 - 1)^2 + (x_2 + 1)^2 = 0.98$ and $C_2 : (x_1 + 1)^2 + (x_2 - 1)^2 = 0.98$, which are represented by 'o's and '+'s, respectively. Points inside each circle represent a class. The points corresponding to C_1 belong to class 1 and the points corresponding to C_2 belong to class 2. Let us try to learn a GP-based binary classifier for this data set. To achieve this, let us consider that every candidate solution in the GP-based system consists of a tree that encodes a discriminant function. When we recursively evaluate a tree $T(\cdot)$ using a data point $\mathbf{x}$, it returns a real value $r_\mathbf{x}^T = T(\mathbf{x})$. If $r_\mathbf{x}^T > 0$, the binary classifier predicts that $\mathbf{x}$ belongs class 1, else it predicts that $\mathbf{x}$ belongs class 2. Let the set of operators (internal nodes) be $\mathcal{F} = \{+, -, \times, \div\}$, and the set of operands (leaf nodes) be $T = \mathbb{R} \bigcup F$, where $F = \{x_1, x_2\}$ is the set of features. We also consider that every operator $f \in \mathcal{F}$ is defined in such a way that if it returns an undefined or infinite value (a value that is beyond storing due to precision limit), the returned value is converted to zero. This conversion may not be the best way to handle this issue. However, every subtree constructed using $\mathcal{F}$ and T must return a real value.

Consequently, each $f \in \mathcal{F}$ is well defined and closed under all possible operands that it may encounter. Hence, this scheme meets the *closure* property. Moreover, for the given problem, we assume that $\mathcal{F}$ satisfies the *sufficiency* property, although strictly speaking this is not true. For example, there are infinite number of equations that can solve the classification problem, and using $\mathcal{F}$ and $\mathcal{T}$, we can design infinitely many trees that can solve the given problem, yet it may not be possible to generate all possible functional form of solutions even for this simple data set. This of course does not cause any problem from a practical point of view as long as $\mathcal{F}$ and $\mathcal{T}$ are able to generate useful solutions at least approximately.

To illustrate further, we show three binary classifiers in Fig. 1, which are denoted by $a : (x_1 - x_2) - 0.1$, $b : (x_1 \times x_2) - 0.1$, and $c : (x_1 \times x_1) - x_2$. Moreover, we show the tree structures corresponding to these three trees respectively in Figs. 2, 3, and 4. As mentioned earlier, infinitely many "correct" solutions to this problem are possible. However, we have purposefully chosen these three particular trees, which have visually similar structures. The solution with the tree a would predict all the points accurately; the solution with the tree b would predict that points of both classes

Fig. 1 An artificial binary classification data set and some binary classifiers

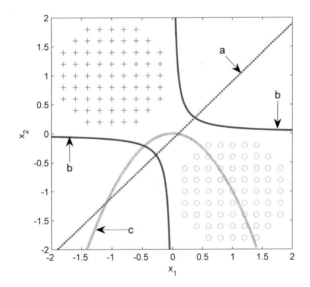

Fig. 2 Tree
$a : (x_1 - x_2) - 0.1$

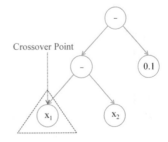

Fig. 3 Tree
$b : (x_1 \times x_2) - 0.1$

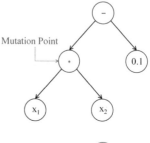

Fig. 4 Tree
$c : (x_1 \times x_1) - x_2$

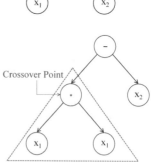

as belonging to class 2; and the solution with tree c would predict some points of class 1 accurately and all points of class 2 accurately.

In any GP-based approach, after the encoding scheme, the second most important issue is the formulation of the objective function that would be used to evaluate the solutions. Here, to evaluate binary classifiers, we use prediction accuracy on the training data set as the evaluation/objective function, this is a simple, straightforward yet an effective strategy. Note that, this objective function is to be maximized.

The third most important issue in the design of a GP-based system is the choice of operators. As we shall see later, there can be many issues like, FS, bloating control, fitness, unfitness, that can be kept in mind while developing these operators. However, here we discuss a primitive crossover and a primitive mutation technique, which are adequate to solve this problem.

The crossover operator requires two parents S_1 and S_2 to generate a new offspring O. To generate the tree of O, it selects two random nodes (may be leaf or non-leaf), one from the tree of S_1 and the other one from the tree of S_2. Let n^{S_1} and n^{S_2} respectively denote those nodes. Then, it replaces the subtree of S_1, which is rooted at n^{S_1}, called the crossover point, by the subtree of S_2, which is rooted at n^{S_2}. To illustrate this with an example, we assume that the trees associated with S_1 and S_2 are respectively c and a. The randomly selected crossover points, and their respective subtrees are also shown in Figs. 2 and 4. After replacing the selected subtree of c (see Fig. 4) by the selected subtree of a (see Fig. 2), the crossover operator generates an offspring O with tree $d : x_1 - x_2$, which is illustrated in Fig. 5. Though, we do not show the tree d in Fig. 1, it is a straight line parallel to tree a and goes through the origin. Moreover, it can also accurately classify the given data set. Thus, though

Fig. 5 Tree $d : x_1 - x_2$

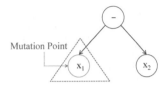

the classifier with tree d yields the same accuracy as that by the classifier with tree a, it is a better choice due to its simplicity (smaller size of the tree).

The mutation operator needs only one solution S. A node from the tree associated with s is randomly selected. If the node is a terminal node, a random number r_n is drawn from [0, 1]. If $r_n < p_{variable}$ ($p_{variable}$ is a pre-decided probability of selecting a variable), the selected terminal node is randomly replaced by a randomly selected variable (feature). Otherwise, the node is replaced by a randomly generated constant $r_c \in R_c$, where $R_c \subseteq \mathbb{R}$. R_c should be chosen judiciously. For example, in this particular example, a good choice of R_c might be $[-2, 2]$. To illustrate the mutation process with an example, let us consider the solution with tree d shown in Fig. 5. Suppose the randomly selected mutation point is the node with feature x_1. Moreover, let us consider that we randomly select to replace this node with a constant node and the constant value (r_c) be 0.01. Then, the mutant tree will be $e : 0.01 - x_2$, which is nothing but a line parallel to the x_1 axis that can also classify the given data set correctly. To illustrate the mutation process, when an internal node is involved, let us consider the solution with tree b (see Fig. 3). In Fig. 3, we have shown the randomly selected mutation point. Let it be replaced with $-$, then it would result tree a, which can classify the problem accurately. However, if this node would have been replaced with $\div$, it would result a new tree $f : (x_1 \div x_2) - 0.1$, which would predict all the points of both the classes.

After defining all necessary components, we can now obtain a GP-based classifier following the steps shown in Algorithm 1. In step 22 of Algorithm 1, it performs the *environmental selection*, which is a necessary component of any GP-based approach. The environmental strategy that we have adopted here is naive. Several criteria and strategies can be adopted for environmental selection. Note that, we also *select* some solutions to perform crossover and mutation respectively in Steps 7 and 15 of Algorithm 1. This selection strategy is called *mating selection*. Instead of randomly selecting these solutions, often some specific criteria are used. For example, since we are maximizing the classification accuracy, a straightforward scheme could be to select solutions using Roulette wheel selection. Another important step, where it is possible to employ a different strategy, is the initialization of the trees. *Ramped half-n-half* method [15] is one of the frequently used methods to initialize GP trees.

We have already mentioned that this algorithm does not explicitly perform any FS and concentrates only on classification accuracy. However, it may implicitly perform FS. To illustrate this, assume that the evolution of this system generates tree $e : 0.01 - x_2$ (we have already mentioned how it can be generated while illustrating the mutation operation). Tree e will have the highest accuracy on the training data,

---

Algorithm 1: A Simple Generational Genetic Programming

---

1 Initialize population $\mathcal{P}$ with N number of solutions.

2 $Generation^{Current} = 0$.

3 **while** $Generation^{Current} < Generation^{Maximum}$ **do**

4      $c_{crossover} = 0$.

5      $\mathcal{O} = \emptyset$.

6   **while** $c_{crossover} < N_{crossover}$ **do**

7   Randomly select two distinct solutions from s_1 and s_2 from $\mathcal{P}$.

8   Generate a new offspring o performing crossover using s_1 and s_2.

9   Evaluate the new offspring o.

10          $\mathcal{O} = \mathcal{O} \bigcup o$.

11          $c_{crossover} = c_{crossover} + 1$

12   **end**

13      $c_{mutation} = 0$.

14   **while** $c_{mutation} < N - N_{crossover}$ **do**

15   Randomly select a solution from s from $\mathcal{P}$.

16   Generate a new offspring o performing mutation on s.

17   Evaluate the new offspring o.

18          $\mathcal{O} = \mathcal{O} \bigcup o$.

19          $c_{mutation} = c_{mutation} + 1$.

20   **end**

21      $\mathcal{U} = \mathcal{P} \bigcup \mathcal{O}$.

22   Select the best N number of solutions from $\mathcal{U}$ and store them in $\mathcal{P}$ (natural or environmental selection).

23      $Generation^{Current} = Generation^{Current} + 1$

24 **end**

25 **return** the best candidate solution s_{best} from $\mathcal{P}$

---

though it uses only one feature (x_2). In this example, the scheme *selected* only one feature (x_2), which has sufficient discriminating power.

2 Genetic Programming for Classification and Feature Selection

There have been several attempts to design classifier using GP [4, 13, 14, 19, 21–23]. A detailed survey on this topic is present in [10]. Some of these methods do not explicitly pay attention to FS [13, 21], while others explicitly try to find useful features to design the classifiers [22, 23]. Some of these GP-based approaches use ensemble concept [19, 23]. In this section, we discuss three existing GP approaches. The first approach [21] introduces a classification strategy employing a single objective GP-based objective search technique. The second approach [22] performs both classification and FS task in an integrated manner. These two schemes use the same multi-tree representation of solutions. On the contrary, the third method [23] decomposes a c-class classification problem into c binary classification problems, and then, performs simultaneous classification and FS. Nevertheless, unlike the first two methods [21, 22], it [23] uses ensembles of genetic programs (binary classifiers) and a negative voting scheme.

2.1 GP-Based Schemes for Feature Selection and Classification

To the best of our knowledge, in [13], the applicability of GP to solve a multi-class classification problem was explored for the first time, where the authors used discriminant function based GP. In [13], the authors decomposed a c-class classification problem to c binary classification problems, and then evolved c genetic programming classifier expressions (GPCEs), one corresponding to each binary classification problem. The GPCE corresponding to the ith class learns to determine the belonging of a point to the ith class. For a given point, a positive response from the GPCE corresponding to the ith class indicates that the point belongs to the ith class. For a point, if multiple GPCEs show positive result, then a special conflict resolution scheme is used. This scheme uses a new measure called *strength of association*. The authors in [13] also introduced an incremental learning scheme. This work showed the capability of GP to automatically discover the features with a better discriminating capability. Later, this work was extended in [14] introducing feature space partitioning, where the feature space was divided into sub-spaces, and then, for every sub-space a GPCE was evolved.

In [19], researchers have proposed a distinct GP based approach for classification of multi-class microarray data sets. Its distinctive nature primarily lies in the structure of the candidate solutions. There [19], every solution that deals with a c-class problem, consists of c sub-ensembles (SEs), where each ensemble posses k trees. In this fashion, every individual comprises of $c \times k$ trees. The outputs of the SEs are decided by a weighted voting scheme that uses the outputs of the k trees of the corresponding SE as arguments and the classification accuracies of the trees (on training data) as the weights. Assuming equal number of training points in each class, they noted that every SE learns with positive to negative ratio $1 : (c - 1)$. To address this data imbalance problem in the fusion of the outputs obtained from the SEs, they furthermore designed a covering score to measure the generalization capability of each SE. For FS, they introduced a measure, called diversity in features (DIF), which estimates the diversity of features in a tree. This measure is used to keep the trees of the same SE to be diverse in terms of their features.

There are two prominent works [27, 28] in the recent literature, in which researchers have reformulated the receiver operating curve convex hull (ROCCH) maximization problem using multi-objective GP to attain binary classification. In both of these works, true positive rate was maximized and false positive rate was minimized simultaneously using an evolutionary multi-objective optimization framework. In [28], they investigated the performances of different evolutionary multi-objective optimization algorithms. In [27], on the other hand the disadvantage of nondominated sorting [8] in EMOAs for ROCCH maximization has been discussed, and to overcome the issues, a new convex hull-based sorting scheme without redundancy has been proposed. This work also introduces an area-based selection scheme, which maximizes the area under the convex hull.

In the past few years, GP based systems [3, 4] have been designed to solve imbalanced binary classification problems. Four new fitness functions have been proposed in [3], which are especially designed for imbalanced binary classification problems. Two of these four intend to enhance the traditional weighted average accuracy measure, whereas, the remaining two are threshold-independent measures, which aim to evolve solutions with good class separability but with faster training times than the area under curve-based functions. To address the data imbalance issue in binary classification, in [4] a multi-objective GP-based scheme has been proposed. This strategy evolves diverse ensembles of genetic programs, which are indeed classifiers with good performance on both the majority and the minority classes. The final ensembles are collections of the evolved nondominated solutions in the population. To ensure diversity in the population, two methods, namely negative correlation learning and pairwise failure crediting, have been proposed.

A GP-based learning technique to evolve compact and accurate fuzzy rule-based classification systems, which is especially suitable for high dimensional problems, has been proposed in [2]. This genetic cooperative-competitive learning approach, where the population constitutes the rule base, learns disjunctive normal form rules. Moreover, a special token competition has been employed to maintain the diversity of the population. It causes rules to compete and cooperate among themselves so that a compact set of fuzzy rules can be obtained. Next, we discuss three methods with reasonable details to explain how GP can be used to design classifiers with or without explicit FS.

2.2 Multi-tree Classifiers Using Genetic Programming [21]

In this approach [21], to solve a c-class classification problem, a GP-based system evolves considering an integrated view of all classes. It uses a multitree representation of the solutions. Some interesting attributes of this approach are: (i) use of a new concept of unfitness, (ii) a new crossover operator, (iii) a new mutation operator, (iv) OR-ing chromosomes of the terminal population, and (v) a weight-based scheme and heuristic rules, which make the classifier capable of saying "don't know" in ambiguous situations. Below we discuss the relevant aspects in details.

Multi-tree Representation of Solutions and Fitness Function: For a c class problem there are c trees in every solution, where each tree corresponds to a class and vice versa. When a data point $\mathbf{x}$ is passed through the ith tree T_i of a given solution, if it produces a positive real value, the tree predicts that $\mathbf{x}$ belong to the ith class. If more than one tree demonstrate positive responses for $\mathbf{x}$, additional methodologies are required to assign a class to **x**. Moreover, the trees are initialized using ramped-half-n-half method with $\mathcal{F} = \{+, -, \times, \div\}$ and $\mathcal{T} = \{$feature variables, $R\}$. Here, R is the set of all possible values in [0,10].

If a tree predicts **x** accurately, the system considers that the accuracy of tree for **x** is one, otherwise zero. The normalized accuracy of all trees of a particular solution for a given set of data points is considered the *fitness* of the solution.

Unfitness: Unlike the concept of *fitness*, *unfitenss* is a less frequently used strategy. However, several works [21–23] have successfully used it to attain better performance. When *fitness* is used in the selection process, more *fit* solutions, i.e., solutions with higher *fitness values* get selected. On the contrary, if *unfitness* is used in the selection process, then more *unfit* solutions, i.e., solutions with higher *unfitness values* get selected. Unfitness-based selection helps unfit solutions to become more fit. For a given problem, a simple choice for unfitness of a tree T_i is the number of training data points, which are wrongly classified by T_i. Let u_j be the unfitness of the jth tree, then this can be easily used to select solutions for genetic operations using Roulette Wheel with probably $u_i/(\sum_{j=1}^{c} u_j)$.

Crossover: At first, τ (tournament size) number of solutions are randomly selected from the population for tournament selection. Then, the best two solutions S_1 and S_2 of the tournament are selected for crossover operation. Suppose the solutions have c trees $T_i^{S_1}$ and $T_i^{S_2}$ $(i = 1, 2, \ldots, c)$ respectively. Now, using the unfitness values of the trees of S_1, suppose the kth tree of S_1, i.e. $T_k^{S_1}$, is selected using Roulette wheel selection. Next, a node from each of $T_k^{S_1}$ and $T_k^{S_2}$ is randomly selected, where the probability of selecting a function node is p_c^f and the same of a terminal node is $(1 - p_c^f)$. After this, the subtrees rooted at the selected nodes are swapped. In addition to this, the trees $T_j^{S_1}$ of S_1 are swapped with the trees $T_j^{S_2}$ of S_2 for all $j > k$.

Mutation: For mutation, a random solution S is selected from the population. Then, with Roulette wheel selection on the unfitness values of its trees, a tree T_k^S is selected. Next, a random node $n^{T_k^S}$ of T_k^S is selected, where the probability of selection of a function node is p_m^f and the same of a terminal node is $(1 - p_m^f)$. If $n^{T_k^S}$ is a function node, it is replaced with a randomly chosen function node. Otherwise, it is replaced with a randomly chosen terminal node. After that, both the mutated and the original trees are evaluated with 50% of the samples of the kth class. Let the fitness values of the two trees be f_m and f_o respectively. If $f_m \geq f_o$, the mutated tree is accepted, otherwise it is retained with a probability of 0.5. If $f_m = f_o$, then both f_m and f_o are evaluated on the remaining 50% samples of the kth class to select one of the two.

Improving Performance with OR-ing: At the end of the evolution process, the best classifier is selected using accuracy. If more than one classifier have the same accuracy, the solution with the smallest number of nodes is selected. However, it may happen that there be two solutions (classifiers) $S_1 = T_i^{S_1}$ and $S_2 = T_i^{S_2}$, $(i = 1, 2, \ldots, c)$, such that, $T_k^{S_1}$ models well for a particular area of the feature space, whereas, $T_i^{S_2}$ performs well in another segment of feature space. To improve the performance, this attribute is exploited as follows. The best solution is OR-ed with all other solutions in the population and the best OR-ed pair is chosen as the final classifier. Note that, here the essence of ensemble of classifiers is introduced, i.e., more than one (two in this case) classifier are used to make the final decision.

2.3 Genetic Programming for Simultaneous Feature Selection and Classifier Design [22]

The GP-based classifier discussed in the last section does not explicitly do FS while designing the classifier. The methodology that we discuss now integrates FS into the classifier design task. This approach [22], simultaneously selects a good subset of features and constructs a classifier using the selected features. Authors here propose two new crossover operators, which carry out the FS process. Like the previously discussed work, here also a multi-tree representation of a classifier is used.

Selection of a Feature Subset Corresponding to Every Solution: Let there be n_f number of features. To generate every solution of the initial population, a subset of features with cardinality r_f ($r_f < n_f$) is randomly selected from the entire feature set, where r_f is selected using Roulette wheel selection with the probability p_{r_f} as follows.

$$p_{r_f} = \frac{n_f - r_f}{\sum_{j=1}^{c} (n_f - j)}. \tag{1}$$

Note that, p_{r_f} decreases linearly with an increase of r_f and $\sum_{r_f=1}^{n_f} p_{r_f} = 1$. After this, r_f features are randomly selected from the entire set of features and then, a single solution is initialized with this selected subset of features.

Fitness: Let f_{raw} be the *raw* fitness of a solution, which is indeed the accuracy achieved by a solution on the entire training data set. Then, the modified fitness function (f_s) that incorporates the FS task in the fitness is defined as follows.

$$f_s = f_{raw}\left(1 + ae^{-\frac{r_f}{n_f}}\right), \tag{2}$$

where

$$a = 2\beta\left(1 - \frac{g_{current}}{g_{maximum}}\right). \tag{3}$$

Here β is a constant, $g_{current}$ is the current generation number, and $g_{maximum}$ is the maximum number of generations of the evolution of GP. Note that, the factor $e^{(-r_f/n_f)}$ is exponentially decreasing with the increase in r_f. Consequently, if two classifiers have the same value of f_{raw}, the one using fewer features would attain a higher fitness value. Moreover, the penalty for using a larger number of features is dynamically increased with generations, which keeps on emphasizing on the FS task with the increase of generations.

Crossover: There are two different crossover strategies: homogeneous crossover (*crossover_hg*) and heterogeneous crossover (*crossover_ht*). *Crossover_hg* restricts the crossover between solutions which use the same feature subset. *Crossover_ht* prefers two solutions for crossover which use more common features. Moreover, the probability of using *crossover_hg* is $p_{hg} = g_{current}/g_{maximum}$, whereas, the probability of using *crossover_ht* is $(1 - p_{hg})$.

To perform *Crossover_ht*, at first τ solutions are randomly chosen as the tournament set. Then, the solution S_1, which has the best f_s among this set, is selected as the first parent. After that, another set with cardinality τ is randomly selected to select the second parent S_2. Let this set be T_{S_2}. Let the chosen subset of a given solution be denoted by a vector $\mathbf{v} = (v_1, v_2, \ldots, v_{n_f})$, such that, if the ith feature is present, then $v_i = 1$, otherwise $v_i = 0$. The similarity measure s_j of the jth solution of T_{S_2} and S_1 is computed as

$$s_j = \frac{\sum_{k=1}^{n_f} v_{S_1 k} v_{jk}}{\max\{\sum_{k=1}^{n_f} v_{S_1 k}, \sum_{k=1}^{n_f} v_{jk}\}}, \tag{4}$$

where, if S_1 uses the kth feature, then $v_{S_1 k} = 1$, otherwise $v_{S_1 k} = 0$; and if the jth solution of T_{S_2} uses the kth feature, then $v_{jk} = 1$, otherwise $v_{jk} = 0$. After that, the jth classifier of the second set is selected with probability p_j^{second} as follows.

$$p_j^{second} = \frac{f_{sj} + \beta s_j}{\sum_{k=1}^{\tau} (f_{sk} + \beta s_k)}, \tag{5}$$

where f_{sk} denotes the fitness (f_s) of the kth solution of T_{S_2} and β is a constant.

In this approach, usually FS is accomplished in the first few generations. Note that, unlike the method that we have discussed in the previous section, this scheme does not use step-wise learning. Because, at the beginning of the evolutionary process, step-wise learning uses a small subset of training samples. Being small in size, it may not be very helpful for the FS process.

2.4 A Multiobjective GP-Based Ensemble for Feature Selection and Classification [23]

Unlike the previous two approaches [21, 22], this approach [23] divides a c-class classification problem into c binary classification problems. Then, it evolves c populations of genetic programs to find out c sets of ensembles, which respectively solve these c binary classification problems. To solve each binary classification problem, a multi-objective archive-based steady-state micro-genetic programming, abbreviated as ASMiGP, is employed. Here to facilitate FS, it exploits the fitness values as well as unfitness values of the features during mutation operation. Both the fitness and the unfitness of features are dynamically altered with generations with a view to obtaining a set of relevant features with low redundancy. A new negative voting strategy is introduced in this context.

The ASMiGP-based Learning Strategy: Each solution in the c populations is a binary classifier, which is encoded by a single tree. When a data point $\mathbf{x}$ is passed through the tree of a binary classifier of the ith population, a positive output and a negative output respectively indicate that $\mathbf{x}$ belongs to the ith class and does not belong to the ith class.

Typically with the evolution of a GP-based system, the average tree size of its population is likely to increase without sufficient or any enhancement in performance of the solutions. This phenomenon is called bloating [29]. Though there are several ways to control bloating [20, 26, 29], one of the prominent ways is to add the tree size as an additional objective. Moreover, if c is sufficiently high, even if the data set is balanced, it may lead to c highly imbalanced binary classification data sets, one for each of the c classification problems. In this case, instead of maximizing classification accuracy or minimizing classification error, simultaneous minimization of false positive (FP) and false negative (FN) might be more suitable. To address these issues, this scheme [23] uses MOGP with the following three objectives: (i)FP, (ii) FN, and (iii) the number of leaf nodes of the tree. The third objective is used to reduce bloating, whereas, the first two objectives incorporate performance parameters of the learning task which help to deal with the imbalance issue. Moreover, to reduce the size of the trees, after generation of any tree throughout the learning phase (using mutation, crossover, or random initialization), the subtrees consisting of only constant leaf nodes is replaced by an equivalent constant leaf node. This also helps to reduce bloating.

This approach [23] uses a special environmental and a special mating selection strategy. It maintains an archive (population) for every binary classification problem. In every generation, it produces a single offspring using either crossover operator or mutation operator. These operations are selected randomly with probability p_c and $(1 - p_c)$ respectively. The female parent for crossover and the only parent for mutation is selected performing Roulette wheel selection on the accuracies of the solutions present in the population. For crossover, the additional (male) parent is selected randomly. After the generation of the new off-spring, it is added to the archive with a special multiobjective archive truncation strategy, which has been adopted from [24, 25]. This environmental selection scheme uses a Pareto-based fitness function, where the Pareto dominance rank is used as the fitness function. This scheme, maintaining a dynamic archive with a hard minimum and a hard maximum size, also ensures diversity in the objective space.

Incorporation of Feature Selection: To assess the discriminating power of a feature, f, the Pearson's correlation, C_f^i, between a vector containing the values of the feature and an ideal feature vector which posses corresponding ideal values for the ith binary classification problem. Specifically, this ideal feature value is unity if the associated training point belongs to the ith class, and zero otherwise [11]. Clearly, a higher value of $|C^i_f|$ designates a stronger discriminative power of feature f for the ith binary classification problem.

For the ith binary classification problem, we intend to incorporate only the features with high discriminating capability from the set off all features $\mathcal{F}_{all}$. To attain this, we assign a fitness and an unfitness value to each of the features, which changes with the evolution. Throughout the process, the features are selected to be added in a tree with probabilities proportional to the current fitness values of the features. On the other hand, features are selected to be replaced (in case of mutation) with probabilities proportional to their current unfitness values. The fitness and unfitness of features corresponding to the ith binary classification problem, during the first 50% evaluations, are defined respectively as in Eqs. (6) and (7).

$$
F^{0\%,i}_{fitness}(f,i) = \begin{cases} \left(\dfrac{C^i_f}{C^i_{max}}\right)^2, & \text{if } \dfrac{|C^i_f|}{C^i_{max}} > 0.3 \\ 0, & \text{otherwise} \end{cases}
\tag{6}
$$

$$
F^{0\%,i}_{unfitness}(f,i) = 1.0 - F^{0\%,i}_{fitness},
\tag{7}
$$

where $C^i_{max} = \max\limits_{f \in \mathcal{F}_{all}} |C^i_f|$. Basically, to eliminate the impact of features with poor discriminating ability during the initial evolution process, Eq. (6) sets their fitness values to zero. Let us assume that $\mathcal{F}_{eval=0\%,i} \subseteq \mathcal{F}_{all}$ is the features with nonzero fitness values.

After completion of 50% evaluations, let the feature subset that is present in the population of the ith binary classification task be $\mathcal{F}_{eval=50\%,i}$. After 50% evaluations, the fitness of all features in $\mathcal{F}_{all} - \mathcal{F}_{eval=50\%,i}$ are set to zero. The assumption behind it is that after 50% evaluations the features which could help the ith binary classification task, would be used by the collection of trees of the corresponding population. Now the fitness and unfitness values of all features in $\mathcal{F}_{eval=50\%,i}$ are changed respectively according to Eqs. (8) and (9).

$$
F^{50\%,i}_{fitness}(f,i) = \begin{cases} \dfrac{|C^i_f|}{\sum\limits_{\substack{f \neq g \\ g \in \mathcal{F}_{50\%,i}}} |\rho_{fg}|}, & \text{if } f \in \mathcal{F}_{eval=50\%,i} \\ 0, & \text{otherwise} \end{cases}
\tag{8}
$$

$$
F^{50\%,i}_{unfitness}(f,i) = e^{-\dfrac{F^{50\%,i}_{fitness}(f,i) - \min_f \{F^{50\%,i}_{fitness}\}}{\max_f \{F^{50\%,i}_{fitness}\} - \min_f \{F^{50\%,i}_{fitness}\}}},
\tag{9}
$$

where ρ_{fg} denotes the Pearson's correlation between feature f and feature g. This helps to select features with high relevance but with reduced level of redundancy, i.e., to achieve maximum relevance and minimum redundancy (MRMR).

After 75% function evaluations, another snapshot of the population is taken. Let the existing features for the ith population be $\mathcal{F}_{eval=75\%,i} \subseteq \mathcal{F}_{eval=50\%,i}$. Then, the fitness and unfitness values of the features in $\mathcal{F}_{eval=75\%,i}$ are altered respectively as in Eqs. (10) and (11).

$$F_{fitness}^{75\%,i}(f,i) = \begin{cases} F_{fitness}^{0\%}(f,i), & \text{if } f \in \mathcal{F}_{eval=75\%,i} \\ 0, & \text{otherwise} \end{cases} \tag{10}$$

$$F_{unfitness}^{75\%,i}(f,i) = 1.0 - F_{fitness}^{75\%,i}(f,i) \tag{11}$$

Crossover: This scheme uses a crossover with male and female differentiation and tries to generate an offspring near the female (acceptor) parent. For this a part of the male (donor) parent replaces a part of the female parent. At first, from each parent a random point (node) is chosen. The probabilities of selecting a terminal node and the non-terminal nodes are respectively p_t^c and $(1 - p_t^c)$. After that, the subtree rooted at the node selected from the female tree is replaced by the subtree rooted at the selected node of the male tree.

Mutation: To mutate a tree, the following operations are performed on the tree: (i) Each constant node of the tree is replaced by a randomly generated constant node with probability p_c^m. (ii) Each function node is visited and the function there is replaced by a randomly selected function with probability p_f^m. (iii) Only one feature node of the tree is replaced by a randomly selected feature node. Among the feature nodes, the mutation point is selected with a probability proportional to unfitness of the features present in the selected tree. Moreover, the feature which is used to replace the old feature, is also selected with a probability proportional to fitness values of the features.

Voting Strategy: After learning is over, a special negative voting strategy is used. At the end of the evolution, c ensembles of genetic programs are obtained: $\mathcal{A} = \{\mathcal{A}_1, \mathcal{A}_2, \ldots, \mathcal{A}_c\}; \forall i, 1 \leq |\mathcal{A}_i| \leq N_{max}$, where c is the number of classes, and $\mathcal{A}_i$ is the ensemble corresponding to the ith class. To determine whether a point $\mathbf{p}$ belongs to the mth class or not, a measure, called *net belongingness* $\mathcal{B}_m^{net}(\mathbf{p})$ corresponding to the mth class for $\mathbf{p}$ is calculated as follows.

$$\mathcal{B}_m^{net}(\mathbf{p}) = \frac{1}{2}\left(\frac{1}{|\mathcal{A}_m|}\sum_{i=1}^{|\mathcal{A}_m|}\mathcal{B}_m^i(\mathbf{p}) + 1.0\right). \tag{12}$$

Here, $\mathcal{B}_m^i(\mathbf{p})$ is defined as

$$
\mathcal{B}_m^i(\mathbf{p}) = \begin{cases} + \left(1.0 - \dfrac{FP_m^i}{FP_m^{max}}\right), & \text{if } \mathcal{A}_m^i(\mathbf{p}) > 0 \\[3mm] - \left(1.0 - \dfrac{FN_m^i}{FN_m^{max}}\right), & \text{otherwise,} \end{cases} \tag{13}
$$

where in Eq. (13), FP_m^i and FN_m^i respectively denote the number of FPs and FNs made by the ith individual of $\mathcal{A}_m$ on the training data set; FP_m^{max} and FN_m^{max} respectively denote the maximum possible FP and the maximum possible FN for the mth class (determined using the training data set); and $\mathcal{A}_m^i(\mathbf{p})$ is the output from ith individual of $\mathcal{A}_m$ for $\mathbf{p}$. Finally, $\mathbf{p}$ is assigned the class k, if $\mathcal{B}_k^{net} = \max\limits_{m=1}^{c}\{\mathcal{B}_m^{net}\}$. Note that, *net belongingness* lies in [0, 1], and a large value of this measure indicates a higher chance of belonging to the corresponding class.

3 Some Remarks

3.1 GP for Classification and Feature Selection

While designing a GP-based system for classification and FS, several important issues should be kept in mind. In this section, we discuss some of these salient issues with details and some ways to address them. Note that, most of these issues depend not only on the GP-based strategy, but also on the data set, i.e, on the number of features, the number of classes, the distribution of the data in different (majority and minority) classes, the feature-to-sample ratio, the distribution of the data in the feature space, etc. Hence, there is no universal rule to handle all of them.

While designing a GP-based system for multi-class classification, the first problem is how to handle the multi-class nature of the data. Primarily there are two ways. The more frequently used scheme is to decompose a c-class problem into c binary classification problems and then develop separate GP-based systems for each of the binary classification problem. Every candidate solution of the ith binary classification system may have one tree, where a positive output from the tree for a given point would indicate that the point belongs to the ith class. To solve every binary classification problem, one may choose to use the best binary classifier (genetic program) found in every binary system, or she may choose to use a set of binary classifiers (ensemble) obtained from these binary systems. If an ensemble-based strategy is used, there need to be a voting (aggregation) scheme. The voting scheme may become more effective if weighted and/or negative voting approach is judiciously designed. Again, to decide the final class label, the outputs (decisions) from every binary system need to be accumulated and processed. Here also, some (weighted, negative) voting scheme, or some especially designed decision making system can be developed. Another comparatively less frequently used strategy is as follows. Every solution consists of a tree, where the output of the tree is divided into c windows using $(c-1)$ thresholds.

Every window in these values would have a one-to-one window-to-class relationship. If the output from a tree falls inside the ith window, the tree would predict that the point belongs to the ith class. The second approach is not recommended as it is less likely to generate satisfactory outcome. Besides, even if it does, the interpretability of the GP-based rules (classifiers) would be low.

If a c-class data set ($c > 2$) is decomposed into c binary classification data sets, even if the class-wise data distribution is near uniform, after decomposition, every binary data set would be imbalanced. The nature of imbalance, in this case, increases with the increase in c. If special care is not taken and only accuracy (or classification error) is used as the objective, each binary module may produce nearly $(c - 1)/c \times 100\%$ accuracy predicting "NO" to every data point. A possible solution to this problem may be to use a multi-objective approach or a weighted sum approach considering both the accuracies on the majority class and the minority class. In our view, multi-objective approach is a better alternative because the choice of weights for the majority and the minority class accuracies is hard to decide, and their choices may have a large impact on the performance of the system.

GP has some interesting issues with generalization of the hidden patterns stored in a particular data set. For example, consider the artificial "XOR" type binary classification data set shown in Fig. 6. Here, uniformly distributed points inside the circles $C_3 : (x_1 - 1)^2 + (x_2 - 1)^2 = 0.75$ and $C_4 : (x_1 + 1)^2 + (x_2 + 1)^2 = 0.75$ belong to class 1, and uniformly distributed points inside the circles $C_5 : (x_1 - 1)^2 + (x_2 + 1)^2 = 0.75$ and $C_6 : (x_1 + 1)^2 + (x_2 - 1)^2 = 0.75$ belong to class 2. The points of class 1 and class 2 have been respectively denoted by the symbols "+" and "o". A solution (binary classifier) with tree $b : x_1 \times x_2 - 0.1$ (see Fig. 3) can accurately classify this data set. Tree b is a small equation and hence a simple model in GP is capable of modeling the hidden pattern stored in this complicated "XOR"

Fig. 6 An artificial "XOR" type binary classification data set and the binary classifier with tree b performing accurately

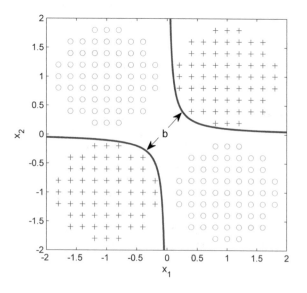

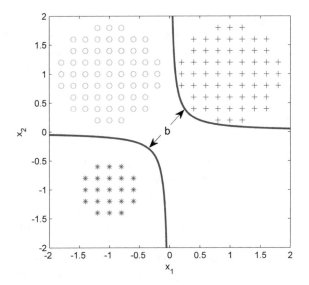

Fig. 7 An artificial binary classification data set and the binary classifier with tree b performing poor generalization

type data set. However, if we try to learn this data set with a multi-layer perceptron type classifier, due to the "XOR" type pattern of the data, it may not be very easy to learn. This example, though, may apparently illustrate GP as a powerful tool, this *specialization* capability of GP may sometimes lead to poor *generalization*. To illustrate a case of poor generalization with an example, let us consider another artificial binary classification data set shown in Fig. 7. There, the class 1 points are denoted using "+" and class 2 points are denoted using "o". The class 1 points are uniformly distributed inside the circle $C_3 : (x_1 - 1)^2 + (x_2 - 1)^2 = 0.75$ and the class 2 points are uniformly distributed inside the circle $C_6 : (x_1 + 1)^2 + (x_2 - 1)^2 = 0.75$. For this data set also, the binary classifier with tree b can accurately classify all the data points. But, for the points denoted by "*" in Fig. 7, the classifier would predict class 1. For these points the classifier should not make any decision. In this example, GP ends up with a poor generalization.

Bloating is another important issue that needs to be addressed to develop a well performing GP-based system. Though there are efficient methods in the literature [20, 26, 29], the following two bloat control strategies are quite straightforward. First, if a single objective approach is used, the size of the tree can be incorporated with the objective function as a penalty factor such that the penalty is minimized. Second, if a multi-objective approach is used, the size of the tree can be added as an additional objective.

As we have already discussed, GP has an intrinsic FS capability. To enforce FS further, whenever a feature is selected to be included in any tree, the model should try to select the best possible feature for that scenario. In a similar fashion, whenever a feature node is removed from any classifier, it should not be an important feature under that scenario. Note that, if an ensemble based strategy is used, for enhanced performance the members of the ensemble should be diverse but accurate.

To exploit this attribute, the model should try to develop member classifiers, i.e., genetic programs, diverse in terms of the features they use. Again, for a single binary classifier, the features used by it should have a controlled level of redundancy but the features should possess maximum relevance for the intended task.

3.2 Parameter Dependency

Any GP-based system requires a set of parameters and the performance of the system is dependant on their chosen values. To discuss the parameter dependency, we select the work proposed in [23]. Note that, this is one of the recent works that performs simultaneous FS and classification. We have chosen this method because it has been empirically shown to be effective on a wide range of data sets with a large number of classes, a large number of features, and a large number of feature to sample ratio. Table 1 shows the parameters and their values used in [23]. We consider CLL-SUB-111 data set used in [23], which is a three class data set with 111 data points and 11,340 features, i.e, the feature-to-sample ratio is 102.16. We have repeated ten-fold cross validation of the method proposed in [23] for ten times. Before training, we have normalized the training data using Z-score, and based on the means and the standard deviations of features in the training data, we also do Z-score normalization of the test data.

Except N_{max} and N_{min}, every other parameter used in [23] is a common parameter required for any GP-based method. While most of the GP-based approaches require a single parameter called population size, the method in [23] requires two special parameters N_{max} and N_{min} to bind the dynamic population size. The performance of any GP-based system is largely dependant on two parameters: the number of function

Table 1 Parameter settings for ASMiGP-based method proposed in [23]

Parameter	Value
Set of functions ($\mathcal{F}$)	$\{+, -, \times, \div\}$
Range of initial values of constants ($\mathcal{C}$)	[0, 2]
Maximum depth of tree during initialization	6
Maximum allowable depth of tree	10
Maximum archive (population) size (N_{max})	50
Minimum archive (population) size (N_{min})	30
Initial probability of feature nodes (p_{var})	0.8
Probability of crossover (p_c)	0.8
Probability of crossover for terminal nodes (p_t^c)	0.2
Probability of mutation for constants (p_c^m)	0.3
Probability of mutation for function nodes (p_f^m)	0.1
Function evaluations for each binary classifier (F_{eval})	400,000

evaluations (F_{eval}) and the population size. Therefore, we choose to show the impact of these parameters on the performance of this method. To attain this, we have repeated our experiment with the CLL-SUB-111 dataset for seven times each with a different parameter setting. The parameter settings used and their corresponding results are shown in Table 2. These results demonstrate that with an increase in F_{eval}, the accuracy increases, and with an increase in the population size (bounded by N_{max} and N_{min}), the number of selected features increases. As indicated by the average tree size in every case the method could find somewhat small trees (binary classifiers). To illustrate this with an example, in Table 3, we have provided six equations, which are generated with the parameter setting S-I. They are the first two (as appeared in the ensemble) binary classifiers of the final populations corresponding to the three binary classification problems associated with the first fold of the first 10-fold cross validation. We have also provided their objective values in that table, which indicate that all of the binary classifiers could produce 100% training accuracy for the corresponding binary classification problem. It is noteworthy that these six rules are simple, concise, and human interpretable.

Table 2 Experimental settings and results

ID	Alteration from Table 1	%TA[a]	FS[b]	TS[c]	(F/T)[d]	%F[e]	$(\%F/T)$[f]
S-I	Unchanged	80.22	342.9	7.33	3.22	3.02	0.03
S-II	$F_{eval} = 400$	67.19	547.8	6.91	2.53	4.83	0.02
S-III	$F_{eval} = 4000$	76.48	334.0	8.57	3.12	2.95	0.03
S-IV	$F_{eval} = 40,000$	77.33	305.9	9.24	3.69	2.70	0.03
S-V	$N_{min} = 90, N_{max} = 150$	80.02	398.4	7.29	3.40	3.51	0.03
S-VI	$N_{min} = 150, N_{max} = 250$	79.03	422.7	7.75	3.68	3.73	0.03
S-VII	$N_{min} = 250, N_{max} = 300$	78.17	436.5	8.39	4.03	3.85	0.04

[a]Test accuracy, [b]Number of features selected per classifier, [c]Tree size, [d]Number of features per tree, [e]Percentage of features selected, [f]Percentage of features selected per tree

Table 3 The first two binary classifiers obtained corresponding to three binary classification problems associated with the first fold of the first 10-fold cross validation

Class	Equation	Objective values
Class 1	$(x_{5261} - 1.3828)$	$(0.0, 0.0, 2.0)$
	$(0.8411 + x_{6911})$	$(0.0, 0.0, 2.0)$
Class 2	$(x_{8962} + x_{9153})$	$(0.0, 0.0, 2.0)$
	$(-1.4391 + x_{5261})$	$(0.0, 0.0, 2.0)$
Class 3	$(x_{8373} - 0.7756)$	$(0.0, 0.0, 2.0)$
	$(x_{8373} - 0.5976)$	$(0.0, 0.0, 2.0)$

4 Conclusion

We have briefly reviewed some of the GP-based approaches to classifier design, some of which do FS. In this context, three approaches are discussed with reasonable details with a view to providing a comprehensive understanding of various aspects related to fitness, unfitness, selection, and genetic operations. We have also briefly discussed the issues related to the choice of parameters and protocols, which can significantly alter the performance of GP-based classifiers. However, there are important issues that we have not discussed. For example, how to design efficiently GP-based classifiers along with feature selection in a big data environment? How to enhance readability of GP classifiers? How to deal with non-numeric data along with numeric ones for designing GP-based classifiers? These are important issues that need extensive investigation.

References

1. http://www.alanturing.net/turing_archive/archive/l/l32/L32-019.html. Accessed 17 Jan 2018
2. Berlanga, F.J., Rivera, A., del Jesús, M.J., Herrera, F.: Gp-coach: genetic programming-based learning of compact and accurate fuzzy rule-based classification systems for high-dimensional problems. Inf. Sci. **180**(8), 1183–1200 (2010)
3. Bhowan, U., Johnston, M., Zhang, M.: Developing new fitness functions in genetic programming for classification with unbalanced data. IEEE Trans. Syst. Man Cybern. Part B: Cybern. **42**(2), 406–421 (2012)
4. Bhowan, U., Johnston, M., Zhang, M., Yao, X.: Evolving diverse ensembles using genetic programming for classification with unbalanced data. IEEE Trans. Evol. Comput. **17**(3), 368–386 (2013)
5. Chakraborty, D., Pal, N.R.: Selecting useful groups of features in a connectionist framework. IEEE Trans. Neural Netw. **19**(3), 381–396 (2008)
6. Colorni, A., Dorigo, M., Maniezzo, V., et al.: Distributed optimization by ant colonies. In: Proceedings of the First European Conference on Artificial Life, vol. 142, pp. 134–142. Paris, France (1991)
7. Cramer, N.L.: A representation for the adaptive generation of simple sequential programs. In: Proceedings of the First International Conference on Genetic Algorithms, pp. 183–187 (1985)
8. Deb, K., Pratap, A., Agarwal, S., Meyarivan, T.: A fast and elitist multiobjective genetic algorithm: Nsga-ii. IEEE Trans. Evol. Comput. **6**(2), 182–197 (2002)
9. Dorigo, M.: Optimization, learning and natural algorithms. Ph. D. thesis, Politecnico di Milano, Italy (1992)
10. Espejo, P.G., Ventura, S., Herrera, F.: A survey on the application of genetic programming to classification. IEEE Trans. Syst. Man Cybern. Part C: Appl. Rev. **40**(2), 121–144 (2010)
11. Hong, J.H., Cho, S.B.: Gene boosting for cancer classification based on gene expression profiles. Pattern Recogn. **42**(9), 1761–1767 (2009)
12. Kennedy, J., Eberhart, R.: Particle swarm optimization. In: Proceedings of International Conference on Neural Networks, vol. 4., pp. 1942–1948. IEEE (Nov 1995)
13. Kishore, J., Patnaik, L.M., Mani, V., Agrawal, V.: Application of genetic programming for multicategory pattern classification. IEEE Trans. Evol. Comput. **4**(3), 242–258 (2000)
14. Kishore, J., Patnaik, L.M., Mani, V., Agrawal, V.: Genetic programming based pattern classification with feature space partitioning. Inf. Sci. **131**(1), 65–86 (2001)

15. Koza, J.R.: Genetic Programming: On the Programming of Computers by Means of Natural Selection. MIT Press, Cambridge (1992)
16. Koza, J.R.: Genetic Programming II: Automatic Discovery of Reusable Programs. MIT press, Cambridge (1994)
17. Koza, J.R., Bennett III, F.H., Stiffelman, O.: Genetic Programming as a Darwinian Invention Machine. Springer, Berlin (1999)
18. Koza, J.R., Keane, M.A., Streeter, M.J., Mydlowec, W., Lanza, G., Yu, J.: Genetic Programming IV: Routine Human-Competitive Machine Intelligence, vol. 5. Springer Science+Business Media (2007)
19. Liu, K.H., Xu, C.G.: A genetic programming-based approach to the classification of multiclass microarray datasets. Bioinformatics 25(3), 331–337 (2009)
20. Luke, S., Panait, L.: A comparison of bloat control methods for genetic programming. Evol. Comput. 14(3), 309–344 (2006)
21. Muni, D.P., Pal, N.R., Das, J.: A novel approach to design classifiers using genetic programming. IEEE Trans. Evol. Comput. 8(2), 183–196 (2004)
22. Muni, D.P., Pal, N.R., Das, J.: Genetic programming for simultaneous feature selection and classifier design. IEEE Trans. Syst. Man Cybern. Part B: Cybern. 36(1), 106–117 (2006)
23. Nag, K., Pal, N.: A multiobjective genetic programming-based ensemble for simultaneous feature selection and classification. IEEE Trans. Cybern. 99, 1–1 (2015)
24. Nag, K., Pal, T., Pal, N.: ASMiGA: an archive-based steady-state micro genetic algorithm. IEEE Trans. Cybern. 45(1), 40–52 (2015)
25. Nag, K., Pal, T.: A new archive based steady state genetic algorithm. In: 2012 IEEE Congress on Evolutionary Computation (CEC), pp. 1–7. IEEE (2012)
26. Poli, R.: A simple but theoretically-motivated method to control bloat in genetic programming. In: Genetic Programming, pp. 204–217. Springer, Berlin (2003)
27. Wang, P., Emmerich, M., Li, R., Tang, K., Baeck, T., Yao, X.: Convex hull-based multi-objective genetic programming for maximizing receiver operating characteristic performance. IEEE Trans. Evol. Comput. 99, 1–1 (2014)
28. Wang, P., Tang, K., Weise, T., Tsang, E., Yao, X.: Multiobjective genetic programming for maximizing roc performance. Neurocomputing 125, 102–118 (2014)
29. Whigham, P.A., Dick, G.: Implicitly controlling bloat in genetic programming. IEEE Trans. Evol. Comput. 14(2), 173–190 (2010)

Genetic Programming for Job Shop Scheduling

Su Nguyen , Mengjie Zhang , Mark Johnston and Kay Chen Tan

Abstract Designing effective scheduling rules or heuristics for a manufacturing system such as job shops is not a trivial task. In the early stage, scheduling experts rely on their experiences to develop dispatching rules and further improve them through trials-and-errors, sometimes with the help of computer simulations. In recent years, automated design approaches have been applied to develop effective dispatching rules for job shop scheduling (JSS). Genetic programming (GP) is currently the most popular approach for this task. The goal of this chapter is to summarise existing studies in this field to provide an overall picture to interested researchers. Then, we demonstrate some recent ideas to enhance the effectiveness of GP for JSS and discuss interesting research topics for future studies.

Keywords Genetic programming · Job shop scheduling · Heuristic

1 Introduction

Scheduling deals with assigning manufacturing resources to process tasks over time. It has a big impact on the cost of operating a manufacturing system, as well as on other performance measures such as on-time delivery. Good scheduling is therefore an

S. Nguyen (✉)
La Trobe University, Bundoora, Australia
e-mail: p.nguyen4@latrobe.edu.au

M. Zhang
Victoria University of Wellington, Wellington, New Zealand
e-mail: mengjie.zhang@ecs.vuw.ac.nz

M. Johnston
University of Worcester, Worcester, UK
e-mail: m.johnston@worc.ac.uk

K. Tan
City University of Hong Kong, Kowloon Tong, Hong Kong
e-mail: kaytan@cityu.edu.hk

© Springer International Publishing AG, part of Springer Nature 2019
J. C. Bansal et al. (eds.), *Evolutionary and Swarm Intelligence Algorithms*, Studies in Computational Intelligence 779,
https://doi.org/10.1007/978-3-319-91341-4_8

143

important factor for a company to be competitive. Because of their complexity, many scheduling problems are considered NP-hard [49], and thus exact methods are usually unable to solve them within a reasonable computational time. Job shop scheduling (JSS) is a good example of such problems in the literature and many heuristics such as dispatching rules [25, 26, 49], tabu search [45], genetic algorithm [8], ant colony optimisation [57], and particle swarm optimisation [54], have been proposed to find quick and acceptable solutions for JSS. Unfortunately, these heuristics are normally problem specific while designing an effective scheduling heuristics is a time-consuming and complicated task.

Dispatching rules is one of the most popular forms of scheduling heuristics that have been investigated since the earliest research on JSS. Basically, dispatching rules aim to prioritise jobs waiting in the shop or in front of a specific machine which then processes jobs based on their assigned priorities (usually jobs with the highest priorities are preferred). Due to their simplicity and ability to react quickly to dynamic changes, dispatching rules have received a lot of attentions from both researchers and practitioners. However, similar to other scheduling heuristics for JSS, designing effective dispatching rules for a particular environment is not a trivial task and involves a lot trials and errors. To handle this issue, different automated approaches have been proposed in the last decade to facilitate the design process. The core of these approaches is machine learning [11, 22] and/or optimisation techniques [6, 18, 24, 37] which aim to learn and search for effective and robust dispatching rules.

Genetic programming (GP) [3, 29] is an evolutionary computation (EC) approach usually used to evolve computer programs for solving a specific task. In recent years, GP has been successfully applied to automatically generate dispatching rules for different scheduling problems. Many different variants of GP such as tree-based GP [29], grammar-based GP [3], and gene expression programming (GEP) [13] are capable of evolving rules that outperform most rules manually designed by human experts in the literature. Some reasons that make GP particularly suitable for these designing tasks are: (1) dispatching rules are usually priority functions [25, 49] which can be easily represented as a mathematical expression by GP trees/programs; (2) GP covers a large (heuristic or program) search space which enables us to discover unknown and effective dispatching rules; (3) representations of GP individuals are generally flexible, which allows sophisticated rules to be encoded and evolved; and (4) GP can take advantages of available EC search mechanisms to enhance the quality of obtained dispatching rules.

Different GP methods have been proposed to deal with JSS, in both static and dynamic environments. The research topics in previous studies are quite diverse and focus on different aspects of GP and JSS, ranging from representations of dispatching rules [6, 37, 43], learning/searching mechanisms [18, 40, 48], fitness evaluations [19, 37], attributes/features analyses [6, 21, 37], and interpretability [6, 18, 21, 37]. The goals of these studies are mainly to: (1) evolve competitive and robust dispatching rules, (2) improve the efficiency of GP for evolving dispatching rules, and (3) satisfy practical requirements in complex environments. Many promising results have been reported in these studies which make GP for JSS an interesting research topic for both operations research and automated heuristic design (hyper-heuristic [7]) communities.

In this chapter, we provide an overview of the current research on GP and JSS in order to help the readers who are interested in this field understand the key aspects, related applications, challenges and opportunities for future studies. The next section will give a brief description of JSS. Section 3 reviews the literature on GP and JSS. In Sect. 4, we revisit some useful ideas that have been proposed to enhance the quality of evolved dispatching rules. Further discussions and conclusions are provided in Sect. 5.

2 Background

The general JSS problem could be simply defined as the scheduling of different jobs to be processed on different machines to satisfy certain objectives. In this case, a job is a sequence of operations, each of which is to be performed on a particular machine. In JSS, the routes of jobs are fixed, but not necessarily the same for each job [49]. An example of a job shop production system is shown in Fig. 1. For the static JSS problem, the shop (or the working/manufacturing environment) includes a set of m machines and n jobs that need to be scheduled. Each job has its own pre-determined route through a sequence of machines to follow and its own processing time at each machine it visits. In static JSS, processing information of all jobs is available. In the dynamic JSS problem, jobs arrive randomly over time and the processing information of jobs is unknown before their arrival.

Over the last few decades, a large number of methods have been developed and applied to JSS, ranging from simple heuristics to artificial intelligence and mathematical optimisation methods. Dispatching rules are perhaps the most straightforward method to deal with both static and dynamic JSS problems [25, 53]. Meanwhile, optimisation is the main research stream to deal with the static JSS problems [49]. A review of these methods is presented in this section. For a broader review of scheduling methods, the readers are encouraged to read Ouelhadj and Petrovic [46] and Potts and Strusevich [52].

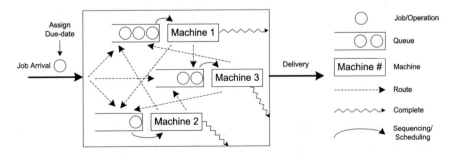

Fig. 1 Job shop scheduling (shop with 3 machines)

2.1 Dispatching Rules

Although there have been many breakthroughs in the developments of exact and approximate methods for JSS; these methods are mainly focused on static problems and simplified job shop environments. General methods like genetic algorithm (GA) can be extended to solve problems with realistic constraints, but the major drawback is its weak computational efficiency. Moreover, as pointed out in [34], the conventional operations research and artificial intelligence methods are often not applicable to the dynamic characteristics of the actual situation because these methods are fundamentally based on static assumptions. For that reason, simple dispatching rules have been used consistently in practice because of their ability to cope with the dynamic changes of the shop.

There have been a large number of rules proposed in the literature and they can be classified into three categories: (1) simple priority rules, which are mainly based on the information related to the jobs; (2) combinations of rules that are implemented depending on the situation that exists on the shop floor; and (3) weighted priority indices which employ more than one piece of information about each job to determine the schedule. Composite dispatching rules (CDR) [25, 26, 49] can also be considered a version of rules based on weighted priority indices, where scheduling information can be combined in more sophisticated ways instead of linear combinations. Pinedo [49] also showed various ways to classify dispatching rules based on the characteristics of these rules. The dispatching rules in this case can be classified as *static* and *dynamic* rules, where dynamic rules are time dependent (e.g. minimum slack) and static rules are not time dependent (e.g. shortest processing time). Another way to categorise these rules is based on the information used by these rules (either local or global information) to make sequencing decisions. A *local* rule only uses the information available at the machine where the job is queued. A *global* rule, on the other hand, may use the information from other machines.

The comparisons of different dispatching rules have been continuously done in many studies [18, 20, 53]. The comparison was usually performed under different characteristics of the shop because it is well-known that the characteristics of the shop can significantly influence the performance of the dispatching rules. Different objective measures were also considered in these studies because they are the natural requirements in real world applications. Although many dispatching rules have been proposed, it is still a challenge for scheduling researchers to develop rules that can perform well on multiple objectives.

2.2 Meta-Heuristic Methods

Since the static JSS is a NP-hard problem [14], finding optimal solutions by mathematical programming methods can be very time-consuming even for reasonable small instances. The research on meta-heuristics for scheduling has been very active in the last two decades, mostly with makespan as the objective. Local search based methods such as simulated annealing [32], large step optimisation [33], tabu search [45], and guided local search [2] have shown very promising results. The focus of these methods is on the development of efficient neighbourhood structures (mainly based on the concept of critical paths and critical blocks) and diversifying strategies to escape from local optima. Since the neighbourhood structures play an important role in these methods, they and their related operators have to be redesigned in order to incorporate real world constraints; even then it is still questionable whether they produce good results.

A more general alternative for solving JSS problems is the use of evolutionary computation methods. GA is one of the most popular methods in this line of research (refer to [8] for a review of GA methods for JSS). More recently, many hybrid algorithms have been proposed to combine the advantages of GA and local search heuristics. Yamada and Nakano [58] presented a GA with multi-step crossover (MSX) for JSS. In this method, MSX was used in combination with a local search heuristic. The preliminary experiments using benchmark instances showed promising performance of the proposed approach. Goncalves et al. [17] proposed a hybrid GA method for JSS to minimise makespan. In this method, the chromosome representation is based on random keys and represents the priorities of operations. An active/non-delay parameter is also applied to restrict the delay time of operations. During the GA search, the schedule is further improved by the neighbourhood local search procedure from [45].

Swarm intelligence methods [4, 5] have also been applied to JSS problems and show very promising results. Sha and Hsu [54] developed a hybrid PSO algorithm (HPSO) that modified the particle position based on preference list-based representation and employed Giffler and Thompson algorithm [16] to decode particle positions into schedules. Moreover, tabu search is also applied to further improve the solution quality. The experimental results showed that HPSO is competitive compared to other meta-heuristics proposed in the literature. Xing et al. [57] proposed a sophisticated ant colony optimisation method in which a knowledge model is used to learn some available knowledge from the optimisation of ACO. The existing knowledge will be used to guide the current heuristic searching. The proposed knowledge-based ant colony optimisation (KBACO) algorithm outperformed some current approaches.

Research on other objectives have also been considered in the literature, especially due date related objectives due to the need to improve the delivery performance in modern manufacturing systems. Pinedo and Singer [50] presented a heuristic to minimise the total weighted tardiness in JSS, which was based on the shifting bottleneck procedure [49] that schedules one machine at a time and used a branching tree to find a good order to schedule the machines. Every node of the tree represents a partial order in which the machines are scheduled. From the experiments, this method yielded solutions that were near optimum on some problems with 10 jobs and 10 machines. Asano and Ohta [1] introduced another heuristic for the minimisation of the total weighted tardiness in JSS that is based on the tree search procedure, also with very promising results. Kreipl [30] proposed an efficient large step random walk (LSRW) method for minimising total weighted tardiness. This method employed different neighbourhood sizes to perform a small step or a large step. The small step consists of iterative improvement while the large step consists of a Metropolis algorithm (similar to the simulated annealing algorithm but with a constant temperature). Essafi et al. [12] proposed a hybrid genetic algorithm which employed an iterative local search and a hybrid scheduling construction procedure to solve this problem. The results showed that the proposed method is very competitive as compared to LSRW [30]. Dispatching rules based meta-heuristics [9, 47, 51] have also been investigated in the literature in order to utilise effective dispatching rules to narrow down the solution search space.

3 Genetic Programming for Job Shop Scheduling

Recently, GP based hyper-heuristics (GPHHs) [7] has become increasingly popular because of their ability to evolve a wide range of program structures corresponding to different types of scheduling rules and heuristics. Another reason that makes GPHHs suitable for this task is that the rules or heuristics evolved by GP can be (partially) interpretable. This is an important characteristic in order to apply obtained heuristics into practical applications. Figure 2 presents the basic framework for evolving dispatching rules with GP. The procedure in this figure starts by preparing training data (datasets or simulation models) for the considered scheduling problems. The initial population of heuristics is then randomly generated (e.g., ramp-half-and-half [3]). In each generation of the evolutionary process, each heuristic is used to provide scheduling decisions for different instances/scenarios in the training datasets or simulation models. The obtained scheduling performance is used to calculate the fitness of each evolved rule. The fitness values obtained by rules in the population decide the chance of each rule to survive and reproduce (with genetic operations) in the next generation. The same routine is applied until the termination condition is met. In the rest of this section, we will review related studies on GP and JSS, which are categorised based on their representations and evolutionary search mechanisms.

3.1 Representations of Dispatching Rules

Representations in GP are very important because they not only decide how evolved rules look like but they also determine how GP can evolve those rules. It is safe to say that most key aspects in GP are governed by the choice of representations. In this section, we will review popular representations employed in the literature and their corresponding genetic operators.

Tree-based representation The most popular representation of dispatching rules is the tree-based representation which is commonly used in the conventional GP, often referred to as tree-based GP (TGP) [3, 29]. This is easy to understand as the tree-based GP is the most established method in the GP literature and the tree-based

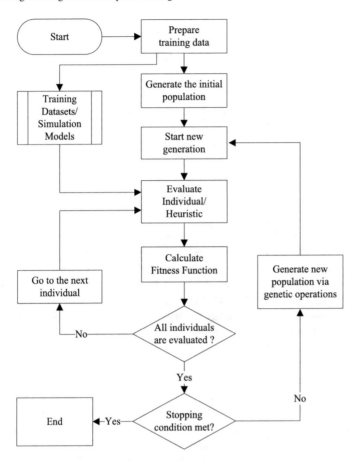

Fig. 2 Evolving dispatching rules with genetic programming

representation can easily be used to represent any (existing) priority functions with appropriate choices of functions and terminals (attributes). An example dispatching rule in the tree structure is shown in Fig. 3. When converted into its mathematical form, it is the same as the critical ratio rule [49] where the terminals t, DD, RT are the decision moment (current time), the due date and remaining processing time of the considered job, respectively. The figure also describes how an evolved rule makes dispatching decisions. When a machine is idle and a new job arrives at the machine, this job will be processed immediately. In the case that a machine has just completed a job and there are still jobs waiting in the queue to be processed at that machine, the dispatching rule will be applied. To assign a priority to a waiting job, the information related to that job will be extracted to be used in the corresponding terminals of the rule. Then, the tree representing the dispatching rule will be evaluated and the output from this evaluation will be assigned to the considered job as its priority (refer to [29] for detailed discussion on how a tree program is evaluated). This procedure will

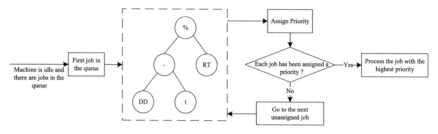

Fig. 3 Tree-based representation of the critical ratio rule $\frac{(DD-t)}{RT}$

be applied until priorities are assigned to all waiting jobs and the job with the highest priority will be processed next.

Due to the complexity of scheduling problems, a single tree may not be sufficient to generate effective and comprehensive scheduling heuristics. Therefore, more sophisticated tree-based representations have been developed. Geiger et al. [15] propose a multiple tree representation to evolve different dispatching rules (trees) for different machines or groups of machines. The goal of this approach is to generate specialised rules to cope with particular requirements of each machine.

In order to create more effective scheduling heuristics for job shops, Jakobovic and Budin [23] present the GP-3 method in which three program trees represent one discriminating function and two priority functions. A special terminal set is used to build the discriminating function which is employed to determine whether the considered machine is a bottleneck. Based on this decision, one of the two priority functions (for bottleneck and non-bottleneck machines) is applied to make scheduling decisions. Nguyen et al. [36] represent the scheduling heuristics by two program trees. The first one is the priority function (the same as previous studies) while the second one represents the look-ahead strategy based on the Giffler and Thompson algorithm [16] to decide how much time idle machines can delay before jobs can be processed. The experiments show that these extended representations can help GPHHs evolve significantly better scheduling heuristics as compared to GPHHs with the single tree representation.

For the tree-based representation, there are many genetic operators available in the GP literature [3]. Subtree crossover and subtree mutation are probably the most commonly used genetic operators in GP to explore new dispatching rules. The subtree crossover creates new individuals for the next generation by randomly recombining subtrees from two selected parents. Meanwhile, the subtree mutation is performed by selecting a node of a chosen individual and replacing the subtree rooted at that node with a newly randomly-generated subtree. Depending on the considered scheduling problems and the structures of the evolved rules, some specialised genetic operators are also applied. For example, Geiger et al. [15] employ restricted crossover and mutation operators to generate dispatching rules for parallel machine scheduling problems. In their approach, only rules from the same machine are allowed to exchange genetic materials. Similarly, Yin et al. [59] also restrict their subtree

Fig. 4 GEP representation
of the rule DD × RT + P

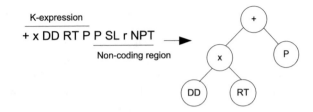

K-expression

+ x DD RT P P SL r NPT

Non-coding region

crossover to be carried out only between the subtrees of similar functions (priority functions and idle-time estimation function).

GEP representation The linear representation of GEP has been applied to construct priority functions [27, 42–44], similar to those evolved by GP with the tree-based representation. GEP genes are also constructed based on the function set and the terminal set. In GEP, the priority function is represented as a chromosome of one or more genes. Each gene in the chromosome represents a fixed length symbolic string which represents a mathematical function. A gene can be divided into two parts: head and tail. While the head can contain both functions and terminals, the tail can only contain terminals. An example GEP chromosome with a single gene is shown in Fig. 4. The gene can be translated into an expression tree by using K-expression. In this example, the first element in the gene + is the root of the tree whose arguments can be obtained from the next elements in the gene. It is noted that the first five elements in the gene have already formed a valid K-expression and the rest of the gene will be ignored in this case. In order to ensure that a valid K-expression can be obtained, the length of the gene will be set such that $t = h(n - 1) + 1$, where h, t, and n are respectively the length of the head, the length of the tail, and the maximum number arguments of a function. In their experiments with the dynamic single machine scheduling problems, Nie et al. [43] show that GEP was very competitive with TGP (better than GP in some cases) and the rules obtained by TGP and GEP were better than all the benchmark heuristics.

Similar to the tree-based representation, the GEP representation can also be extended to cope with multiple scheduling decisions. Nie et al. [42, 44] develop GEP methods to deal with flexible job shop scheduling problems. In their methods, each GEP individual contains two chromosomes for making sequencing and routing decisions or a chromosome will contain two genes representing the two scheduling rules. The results show that the new GEP methods can evolve scheduling heuristics that outperform heuristics in the literature and the GEP method that deals with a single scheduling decision.

In order to evolve more sophisticated rules, multiple genes can also be used to represent multiple functions which can be combined by using a simple summation of these functions [43] or explicitly using a control gene to combine the outputs from these functions [44]. In the latter approach, the control gene is a dedicated gene which is used to characterise the relationship between outputs obtained from other genes. The control gene used the same function set as other genes and the terminal set consisting of outputs from other genes. While this representation can help GEP evolve more sophisticated rules, it will also increase the computational time as well as the search space of GEP.

The genetic operators in GEP can be considered as hybrids between those of genetic algorithm (GA) and TGP. The subtree crossover and subtree mutation from TGP can also be applied to GEP. However, because of the difference in data structure (linear vs tree), GEP needs to explicitly transverse through elements in a gene to identify the subtree. Because the length of a GEP gene is fixed, the same genetic operators such as the point mutation and the one-point/two-point crossover in GA can also be applied [42, 44]. Special transposition operators are also employed in GEP to randomly select a fragment of the chromosome and insert it into the head.

Grammar-based representation Different from the tree-based representations which mainly focus on evolving priority functions (dispatching rules), grammar-based representations are usually used to construct high-level heuristics composed of several low-level heuristics and solution attributes. Although grammar-based GP has been developed to evolve heuristics for many hard combinatorial problems, their applications in manufacturing scheduling are still very limited. Nguyen et al. [37] develop a grammar-based representation for GP to evolve adaptive dispatching rules for job shop scheduling. The heuristics evolved with this representation is quite similar to decisions trees which try to find out which (available) candidate dispatching rule should be applied and what non-delay factor should be used given some specific machine/system status. The advantage of this representation is that the obtained rules can be interpreted easier as compared to evolved priority functions previously discussed. Also, by using a set of candidate rules which have been readily coded, the evaluations of these rules are faster than those of rules with the tree-based representation. On the other hand, the disadvantage of this representation is that it depends a lot on the available problem-domain knowledge to choose appropriate machine/system attributes and candidate rules. If the candidate rules cannot cover all situations, the evolved adaptive rules may not provide satisfactory results.

3.2 Search Mechanisms

The traditional search mechanism in Fig. 2 is currently the most common technique to deal with scheduling problems. Regardless of its simplicity, this framework is able to discover very effective scheduling heuristics. However, in order to deal with more complicated design issues such as multiple scheduling decisions and multiple objectives, specialised search mechanisms will be needed.

Evolutionary multi-objective optimisation Multiple conflicting objectives are a natural characteristic in real world applications and the design of new scheduling heuristics also need to consider this issue. One advantage of using GPHHs for designing heuristics is that their search mechanisms are very flexible and many advanced techniques have been developed to cope with multiple objectives.

Tay and Ho [55] aim to tackle three objectives (makespan, mean tardiness, and mean flowtime) when using GP to evolve dispatching rules for a flexible job shop. In order to simplify the design problem, the three objectives are aggregated by using the weighted sum approach with the same weight for each objective. However, because the scale of each objective as well as the knowledge about the objective search space is unknown, this approach can lead to unsatisfactory results. For that reason, the rules evolved by their GP method are sometimes worse than simple rules such as FIFO [55]. When these evolved rules are examined in a long simulation [18], they are only slightly better than the earliest release date (ERD) rule and worse than the shortest processing time (SPT) rule with respect to mean tardiness. This suggests that using the weighted aggregated objective to deal with multi-objective design problem is not a good approach if the prior knowledge about the objective functions is not available.

Nguyen et al. [38] develop a multi-objective genetic programming based hyper-heuristic (MO-GPHH) for dynamic job shop scheduling. In this work, they aim to evolve a set of non-dominated dispatching rules for five common objective functions in the literature. By relying on the Pareto dominance rather than any specific objective, the proposed MO-GPHH was able to evolve very competitive rules as compared to existing benchmark rules in the literature. Their results show that it is very easy for MO-GPHH to find rules that totally dominate simple rules such as FIFO and SPT regarding all five considered objectives. The proposed MO-GPHH can also find rules that dominate more sophisticated rules such as apparent tardiness cost (ATC), RR, 2PT+WINQ+NPT, and cost over time (COVERT) [25, 26, 53] in most of its runs. The analyses show that evolving the Pareto front is more beneficial as compared to evolving a single rule as many unknown and helpful trade-offs are be discovered. Similar methods have been applied to evolve comprehensive scheduling policies for dynamic job shop scheduling [40] and order acceptance and scheduling [39] and showed promising results.

Coevolution The advantages of automatic design of scheduling heuristics have been demonstrated in the previous study. However, the drawback of this approach is the high computational time. Even though the obtained heuristics are very fast, training hundreds or thousands of these heuristics under different training examples can be very time consuming. To cope with this issue, GP can evolve heuristics in parallel to reduce the computational times and hopefully the complexity of the problem.

Miyashita [35] proposes three multi-agent learning structures based on GP to evolve dispatching rules for JSS. The first one is a homogeneous agent model which is basically the same as other GP methods which evolves a single dispatching rule for all machines. The second model treated each machine (resource) as a unique agent which requires distinct heuristics to prioritise jobs in the queue. In this case, each agent has its own population to evolve heuristics with GP. Finally, this research proposed a mixed agent model in which resources are grouped based on their status. Two types of agents in this model are bottleneck agent and non-bottleneck agent. Because of the strong interactions between agents, credit assignment is difficult. Therefore, the performance of each agent is directly measured by the quality of the entire schedule. The experimental results show that the distinct model has better training performance compared to the homogeneous model. However, the distinct model has overfitting issues because of the too specialised rules (for single/local machines). The mixed agent model shows the best performance among the three when tested under two different shop conditions. The drawback of this model is that it depends on some prior-knowledge (i.e. bottleneck machines) of the job shop environment, which can be changed in dynamic situations.

To deal with multiple scheduling decisions (sequencing and due date assignment) in job shops, Nguyen et al. [40] develop a GP based cooperative coevolution approach in which scheduling rules for a scheduling decision are evolved in their own subpopulation. Similar to [35], the fitness of each rule is measured by the overall performance obtained through cooperation. Specialised crossover, archiving and representation strategies are also developed in this study to evolve the Pareto front of non-dominated scheduling heuristics. The results show that the cooperative coevolution approach is very competitive with some other evolutionary multi-objective optimisation approaches. The analysis also indicates that the proposed cooperative coevolution approach can generate more diverse sets of non-dominated scheduling heuristics.

Multi-stage learning/optimising In order to further utilise the outputs of GPHHs for enhancing scheduling performance, some multi-stage learning/optimising approaches have been proposed. Kuczapski et al. [31] develop two hyper-heuristics to generate initial populations of GA for JSS. The first hyper-heuristic uses GP to evolve composite dispatching rules similar to previous studies [10, 18, 55]. The second one tries to find composite rules through simple weighted sum of priorities generated by some existing dispatching rules. Another GA is used in this GPHH to search for the weight of each existing dispatching rule. The comparison showed that rules generated by GA is better than those generated by GP. A reason for the poor performance of GP in this case may be that the population size (20) is too small and does not provide GP with enough genetic materials to synthesise effective rules. The experimental results showed that the two hyper-heuristics can find rules to generate good initial solutions for JSS and significantly enhance the performance of GA, particularly improving the quality of solutions up to 10% and reducing the computational time up 50% [31]. One drawback of this approach is that the proposed GPHHs have to be applied for each instance and reusability of evolved dispatching rules has not been investigated.

The multi-stage approach is not only applied to static scheduling problems but also to dynamic scheduling problems. Pickardt et al. [48] propose a two-stage approach to evolving dispatching rule sets for semiconductor manufacturing. In the first stage, GP is used to evolve general dispatching rules. The best obtained dispatching rule is combined with a list of benchmark dispatching rules to generate a set of candidate rules. In the second stage, a $\mu + \lambda$ evolutionary algorithm (EA) [48] is used to select the most suitable dispatching rule in the set of candidate rules for each work centre in the shop. The experiments in this paper compare the performance of the two-stage hyper-heuristics with the pure GP and EA hyper-heuristics. The results show that the three hyper-heuristics outperformed benchmark dispatching rules and the two-stage hyper-heuristics produced significantly better performance than the other two hyper-heuristics.

4 Performance Enhancement Revisited

The previous section has shown some successful applications of GP for JSS. Several clever ideas have been introduced in previous studies that can be used to enhance the performance of GP for JSS. Although many ideas are proposed for specific scheduling problems (e.g., flexible job shops with routing and sequencing decisions), some ideas are quite generic and can be easily applied to enhance the quality of GP for JSS. In this section, we will revisit these generic ideas to demonstrate how they can be applied and their usefulness. All experiments are conducted using the same GP system and simulation environments. Different performance measures of JSS are examined to verify the general effectiveness of the considered ideas.

4.1 Experimental Settings

All experiments in this section are based on the simulation model of a symmetrical job shop which has been used in previous studies on dispatching rules [6, 25, 38]. Here are the simulation configuration:

- 10 machines
- Each job has 2–14 operations (re-entry is allowed)
- Processing time follows discrete uniform distribution $U[1, 99]$
- Job arrivals follow Poisson process
- Due date = current time + allowance factor × total processing time (allowance factor of 4 is used in our experiments)
- Utilisation of the shop is 95%
- No machine break-down; preemption is not allowed
- Weights of jobs are assigned based on the 4:2:1 rule [30, 50].

In each simulation replication, we start with an empty shop and the interval from the beginning of the simulation until the arrival of the 500th job is considered as the warm-up time and the statistics from the next completed 2000 jobs [20] will be used to calculate performance measures. Three scheduling performance measures examined in our experiments are: (1) mean flowtime, (2) mean tardiness, (3) total weighted tardiness. Although this simulation model are relatively simple, it still reflects key issues of real manufacturing systems such as dynamic changes and complex job flows. This section only considers the shop with high utilisation (95%) and tight due date (allowance factor of 4) because scheduling in this scenario is more challenging, which is easier to demonstrate the usefulness of GP. Table 1 shows the terminal set and function set used in our experiments. The parameters used in GP are presented in Table 2. The results for each GP method in this section are based on 30 independent runs.

Table 1 Terminal and function sets of GP

Symbol	Description
rJ	Job release time (arrival time)
RJ	Operation ready time
RO	Number of remaining operation within the job.
RT	Work remaining of the job
PR	Operation processing time
DD	Due date of the job
RM	Machine ready time
SL	Slack of the job $= DD - (t + RT)$
WT	Is the current waiting time of the job $= \max(0, t - RJ)$
#	Random number from 0 to 1
NPR	Processing time of the next operation
WINQ	Work in the next queue
APR	Average operation processing time of jobs in the queue
Function set	$+, -, \times, \%$, min, max

*t is the time when the sequencing decision is made

Table 2 Parameter settings

Parameter	Description
Initialisation	Ramped-half-and-half
Crossover rate	80%
Mutation rate	15%
Reproduction rate	5%
Maximum depth	8
Number of generations	100
Population size	500
Selection	Tournament selection (size $= 5$)

4.2 Training Simulation Replications

Using discrete event simulation is a conventional and suitable approach to assess the performance of dispatching rules. In order to reliably measure the effectiveness of evolved rules, a large number of simulation replications are needed (e.g., 30–50 simulation replications are usually needed to accurately estimate the performance of rules in the scenario described in Sect. 4.1). However, using simulation to evaluate fitness of evolved rules is also the most time-consuming part in GP for JSS. Therefore, only a small number of replications are usually used for fitness evaluations during the evolution (training) process.

How to effectively use the computational budget, i.e. total number of simulation replications, is an interesting and important research question. Hildebrandt et al.

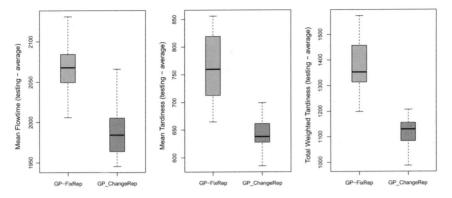

Fig. 5 *Fixed* simulation replications versus *changing* simulation replications

[18] has investigated different trade-offs between numbers of training replications and maximum numbers of generations when trying to evolve dispatching rules for a semiconductor manufacturing systems. From their analyses, the best setting is to use only one replication to evaluate fitness of rule; however, different replications (different random seeds for the simulator) should be used in different generations. In their method, the best rule obtained from each generation is fully evaluated to validate its real performance. The experimental results of *fixed* simulation replication and *changing* simulation replication strategies for our scenario are shown in Fig. 5 (for presentation purpose, the total weighted tardiness is normalised by dividing it by the number of jobs). The values in these boxplots are the average performance measures over 50 simulation replications.

In this experiment, GP-FixRep uses only two fixed simulation replications during the evolution (thus, only 50 generations are performed for a fair comparison) and GP-ChangeRep changes the simulation replication when moving to the next generation). From the experiments, it is quite clear that changing simulation replications can help GP evolve more effective dispatching rules regardless of performance measures. As changing simulation replications can be easily implemented, using this strategy is a good way to enhance the quality of evolved rules when dealing with a single objective (i.e., performance measure), especially when the computational times are restricted or the simulation is computationally too expensive.

4.3 Smooth and Nonsmooth Evolved Dispatching Rules

Selecting an appropriate function set is important in GP as it decides the search space of evolved rules. Since most popular rules in the literature are relatively simple and can be easily constructed by using basic arithmetic operators ($+$, $-$, $\times$, %), it is unclear if using more sophisticated functions such as *if*, max, min is necessary. In this section, we examine GP with two different function sets (1) GP-Smooth where

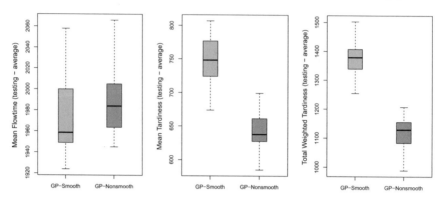

Fig. 6 *Smooth* versus *nonsmooth* evolved dispatching rules

only basic arithmetic operators are used and (2) GP-Nonsmooth where arithmetic operators and max, min are used. The comparison of the two settings are shown in Fig. 6.

In the case with mean flowtime as the performance measure, there is no significant difference between the two settings and GP-Smooth is even slightly better than GP-Nonsmooth. It is understandable because most effective existing rules in the literature for minimising flowtime are usually simple (usually linear combinations of different attributes) such as shortest processing time (SPT), or 2PT+WINQ+NPT [20]. Therefore, min, max seems to be redundant here and may deteriorate the performance of GP.

For minimising mean tardiness and total weighted tardiness, GP-Nonsmooth convincingly outperform GP-Smooth. These results suggest that basic arithmetic operators are not sufficient to construct rules to deal with complex scheduling problems (e.g., complex performance measures). The smooth evolved rules generated by GP-Smooth cannot possess complex behaviours to cope with different situations in the simulation. As we examine the best evolved rules from GP-Nonsmooth, it is easy to see that max, min functions create a complex behaviour for evolved rules which cannot be easily created using arithmetic operators. However, it is noted that we also try to avoid too complex functions such if or max, min with multiple arguments as they can significantly increase the search space without making any real contributions. From our past experiments, max, min along with basic arithmetic operators are reasonably sufficient to generate very complex (nonsmooth) rules. These issues also need to be taken into account when developing new representations for dispatching rules.

4.4 Surrogate Model

Surrogate models have been employed in many EC applications, especially when the fitness evaluations are expensive. However, compared to traditional EC methods, using surrogate models in GP is a lot more complicated. Approximating the fitness of GP individuals is challenging because it is difficult to capture the behaviours of evolved programs, or dispatching rules in our applications.

Hildebrandt and Branke [19] investigated two surrogate models for evolving dispatching rules to minimise mean flowtime. In their approach, the fitness of an evolved rule is approximated by using the fitness of the most similar rules generated in the previous generations. Based on this idea, two similarity measures are proposed. The first measure is based on the genotype similarity of evolved rules, i.e. the similarity in their structure. The second measure is calculated based on the similarity of evolved rules, i.e the similarity in the way they prioritise jobs. Their (simplified) proposed surrogate-assisted GP (SGP) can be summarised as follows:

  i. Initialise the population with N rules
 ii. Evaluate and determine (real) fitness for all evolved rules
iii. Apply genetic operator to generate $m \times N$ new rules
 iv. Approximate the fitness of newly generated rules
  v. Put rules with the best fitness into the population of the next generation
 vi. Stop if the termination condition is met; otherwise, back to step (ii).

SGP in [19] used fixed simulation replications and utilised individuals in the last two generations to approximate fitness of newly generated rules. For the surrogate model, the behaviour of an evolved rule is characterised by a decision vector based on a reference rule (2PT+WINQ+NPT) and the similarity of two rules is measured by the distances of their corresponding decision vectors (see [19] for a detailed description). In our experiments, we further examine the performance of SGP with changing simulation replications and different reference rules. Different from [19], because fitness of rules in different generations is not compatible (as different replications are used), we only use rules in the most recent generation for fitness approximation. The results of our experiments are shown in Fig. 7.

In Fig. 7, GP represents simple GP method with changing replications and the extended function set. SGP-FIFO and SGP-REF are our SGP versions with FIFO and

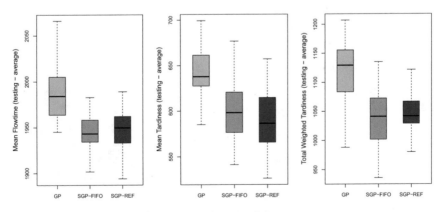

Fig. 7 Performance of *surrogate-assisted* GP methods for JSS

2PT+WINQ+NPT as reference rules respectively. The results show that surrogate approaches are more effective as compared to the simple GP approach when the same computational budget is used. It is noted that SGP-FIFO and SGP-REF are slightly slower than GP because of fitness approximation. However, as the complexity of scheduling environments increases, the time for fitness approximation will be negligible as compared to the time for full fitness evaluations. More detailed analyses of SGP are presented in Figs. 8, 9 and 10.

In each figure, the left part shows the progress of the best performance measure (based on 50 simulation replications) across generations (averaged over 30 independent runs). The right part shows the ratio between the best fitness and average fitness in each generation. The behaviour in the right part is not stable because the fitness function changes across generations. For all three performance measures, it is easy to see that SGP methods can find good rules a lot faster than GP. In three cases, SGP methods can find the best rules evolved by GP by using only 50 generations (two times faster than GP). Although, SGP methods employ elitism in their selection, they did not suffer from premature convergence. For the right parts of Figs. 8, 9 and

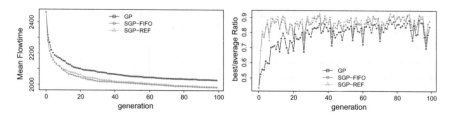

Fig. 8 Behaviours of *surrogate-assisted* GP – minimise mean flowtime

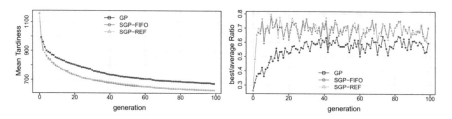

Fig. 9 Behaviours of *surrogate-assisted* GP – minimise mean tardiness

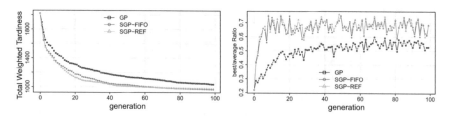

Fig. 10 Behaviours of *surrogate-assisted* GP – minimise total weighted tardiness

10, it is obvious that SGP can also find better rules for each training replication as compared to GP. This helps confirm that surrogate model is still be very useful even when the simulation replication changes across generations. Therefore, we can take advantages of both changing simulation replications and surrogate models.

4.5 Multi-objective

One big advantages of using GP or EC for automated design of dispatching rules is that evolutionary multi-objective optimisation (EMO) has been mature over the last decade. Many effective and efficient EMO techniques have been developed to deal with hard multi-objective problems. Therefore, there is no reason not to utilise EMO to find the set of non-dominated rules, which help us understand better about possible trade-offs before selecting appropriate dispatching rules to apply. Figure 11 demonstrates the usefulness of evolving non-dominated rules. This figure presents the non-dominated rules for five performance measures found by MO-GPHH [41] for our considered scenario. The black circles represent evolved non-dominated rules while the red crosses represent 31 rules developed in the literature. It is not hard to see that there are many interesting trade-offs that have been ignored in the literature. For example, we can find very effective rules to minimise both maximum flowtime (Fmax) and percentage of tardy jobs (%T); however, these rules have never been discovered.

One problem with evolving heuristics for multi-objective problems is overfitting. The chance of creating overfitted heuristics through crossover and mutation is actually quite high when Pareto dominance is used as the criteria for individual selection in GPHHs. A heuristic in the Pareto front accidentally generated may not be dominated by other good non-dominated heuristics, especially when many objectives are considered, even though the heuristic can contain many useless components. This issue also occurs in single objective methods; however, it is less severe because only one objective is considered and later generation can find more compact heuristics to replace the unnecessarily lengthy heuristics. It would be interesting to further extend the performance enhancement strategies mentioned in previous sections to improve the quality of evolved non-dominated rules.

4.6 Simplification

Evolving more compact dispatching rules is important in GP for JSS. More compact rules are easier to understand, especially when rules are in the form of priority functions. Moreover, compact rules will made the fitness evaluations faster and will significantly reduce the computational times of GP. In this section, we try to perform online simplification to hopefully help GP explore more compact rules through the evolution process. Because we are dealing with mathematical expression, a straight-

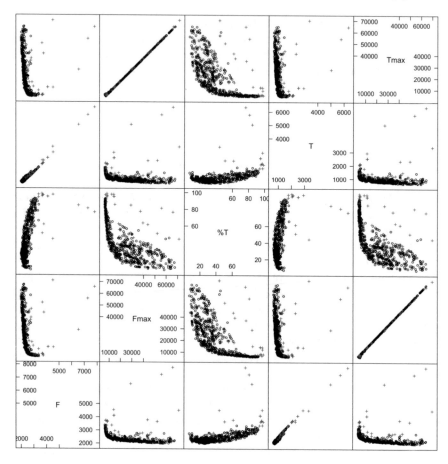

○ Evolved rules + Existing rules

Fig. 11 Distribution of rules on the evolved Pareto front

forward simplification technique is to apply simple symbolic simplification rules
[28, 56].

In our experiments, we examine three methods to incorporate online simplification into GP. GPSimFit only simplifies evolved rules for fitness evaluations and the
original rules are returned for genetic operations. GPSimALL simplifies all generated
rules in the population. Finally, GPSimHalf randomly simplifies a half of rules in the
population. The results from the three simplification strategies and the simple GP
method are shown in Fig. 12. Basically, there is no significant difference between the
three methods in term of performance measures. Simplification also does not ensure
that more compact rules will be obtained at the end of GP runs. When examining the
rules during the evolution process, we observe that simplification only make evolved
rules compact temporarily. Over time, more complex rules are still generated with

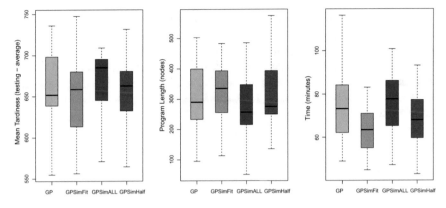

Fig. 12 Simplification in GP

subtrees that cannot be further simplified. Therefore, the final rules are still quite complex. The only strategy with some potential benefit is GPSimFit where the computational times can be significantly reduced when times for fitness evaluations are reduced due to simplification. Since the original building blocks are preserved in this case, simplification is still useful even at the latter stage of the evolution.

5 Conclusions

In this chapter, we have provided an overview of current research on GP and JSS. Although JSS has been studied for decades, automated design of dispatching rules is still a relatively new research direction. The use of GP for evolving dispatching rules has made the design process a lot easier and more productive. Instead of spending time fabricating or improving dispatching rules, scheduling experts can focus more on investigating behaviours of obtained rules, comparing the trade-offs between different rules, and handling real-world constraints.

In future studies, representations of dispatching rules are still a key research topic to further enhance the effectiveness of GP for JSS. Since practical manufacturing systems can be very complex, many aspects need to be considered. Smart function and feature/attribute selection is worth investigating to make GP search more effective. Multiple scheduling decisions and multiple objectives are interesting research topics; however, they are still challenges for GP in order to effectively handle these issues. Surrogate-assisted GP is promising and still has a lot of room for further improvements. Finally, interpretability of rules needs to receive more attentions. Evolved rules are still quite complicated and we still have to manually simplify and transform evolved rules to an understandable forms. Therefore, there is a need of more systematic and effective approaches (e.g., visualisation, automatic analyses) to help us interpret evolved rules. This is also a key issue to gain the confidence of users on GP systems.

Because there are many optimisation problems which share similar characteristics of job shop scheduling, advances gained from this field can also be applied to other applications, especially ones with sequencing related decisions. The combination of GP with other optimisation methods should be an interesting topic to explore in future studies.

References

1. Asano, M., Ohta, H.: A heuristic for job shop scheduling to minimize total weighted tardiness. Comput. Ind. Eng. **42**, 137–147 (2002)
2. Balas, E., Vazacopoulos, A.: Guided local search with shifting bottleneck for job shop scheduling. Manage. Sci. **44**, 262–275 (1998)
3. Banzhaf, W., Nordin, P., Keller, R., Francone, F.: Genetic Programming: An Introduction. Morgan Kaufmann, San Francisco (1998)
4. Beni, G., Wang, J.: Swarm intelligence in cellular robotic systems. In: Dario, P., Sandini, G., Aebischer, P. (eds.) Robots and Biological Systems: Towards a New Bionics? NATO ASI Series, vol. 102, pp. 703–712. Springer, Berlin, Heidelberg (1993). https://doi.org/10.1007/978-3-642-58069-7_38
5. Bonabeau, E., Dorigo, M., Theraulaz, G.: Swarm Intelligence: From Natural to Artificial Systems. Oxford University Press, Inc., New York, NY, USA (1999). http://portal.acm.org/citation.cfm?id=328320
6. Branke, J., Hildebrandt, T., Scholz-Reiter, B.: Hyper-heuristic evolution of dispatching rules: a comparison of rule representations. Evol. Comput. (2014) (in press). (https://doi.org/10.1162/EVCO_a_00131)
7. Burke, E.K., Hyde, M.R., Kendall, G., Ochoa, G., Ozcan, E., Woodward, J.R.: Exploring hyper-heuristic methodologies with genetic programming. In: Mumford, C., Jain, L. (eds.) Computational Intelligence, Intelligent Systems Reference Library, vol. 1, pp. 177–201. Springer, Berlin, Heidelberg (2009)
8. Cheng, V.H.L., Crawford, L.S., Menon, P.K.: Air traffic control using genetic search techniques. In: McClamroch, N.H., Sano, A., Gruebel, G. (eds.) In: Proceedings of the 1999 IEEE International Conference on Control Applications, vol. 1, pp. 249–254. IEEE Press, Piscataway, NJ (1999)
9. Chiang, T.C., Shen, Y.S., Fu, L.C.: A new paradigm for rule-based scheduling in the wafer probe centre. Int. J. Prod. Res. **46**(15), 4111–4133 (2008)
10. Dimopoulos, C., Zalzala, A.M.S.: Investigating the use of genetic programming for a classic one-machine scheduling problem. Adv. Eng. Softw. **32**(6), 489–498 (2001)
11. El-Bouri, A., Balakrishnan, S., Popplewell, N.: Sequencing jobs on a single machine: a neural network approach. Eur. J. Oper. Res. **126**(3), 474–490 (2000)
12. Essafi, I., Mati, Y., Dauzère-Pérès, S.: A genetic local search algorithm for minimizing total weighted tardiness in the job-shop scheduling problem. Comput. Oper. Res. **35**(8), 2599–2616 (2008)
13. Ferreira, C.: Gene Expression Programming: Mathematical Modeling by an Artificial Intelligence, 2nd edn. Springer, Germany (2006)
14. Garey, M.R., Johnson, D.S., Sethi, R.: The complexity of flowshop and jobshop scheduling. Math. Oper. Res. **1**(2), 117–129 (1976)
15. Geiger, C.D., Uzsoy, R., Aytuğ, H.: Rapid modeling and discovery of priority dispatching rules: an autonomous learning approach. J. Sched. **9**(1), 7–34 (2006)

16. Giffler, B., Thompson, G.L.: Algorithms for solving production-scheduling problems. Oper. Res. **8**(4), 487–503 (1960)
17. Goncalves, J.F., de Magalhaes Mendes, J.J., Resende, M.G.C.: A hybrid genetic algorithm for the job shop scheduling problem. Eur. J. Oper. Res. **167**(1), 77–95 (2005)
18. Hildebrandt, T., Heger, J., Scholz-Reiter, B.: Towards improved dispatching rules for complex shop floor scenarios—a genetic programming approach. In: Pelikan, M., Branke, J. (eds.) In: GECCO'10: Proceedings of the 12th Annual Conference on Genetic and Evolutionary Computation, pp. 257–264. ACM Press, New York (2010)
19. Hildebrandt, T., Branke, J.: On using surrogates with genetic programming. Technical Report, Warwick Business School (2014)
20. Holthaus, O., Rajendran, C.: Efficient jobshop dispatching rules: further developments. Prod. Plann. Control **11**(2), 171–178 (2000)
21. Hunt, R., Johnston, M., Zhang, M.: Evolving "less-myopic" scheduling rules for dynamic job shop scheduling with genetic programming. In: GECCO'14: Proceedings of Genetic and Evolutionary Computation Conference (2014) (to appear)
22. Ingimundardottir, H., Runarsson, T.P.: Supervised learning linear priority dispatch rules for job-shop scheduling. In: Coello Coello, C.A. (ed.) Learning and Intelligent Optimization, LNCS, vol. 6683, pp. 263–277. Springer, Berlin, Heidelberg (2011)
23. Jakobović, D., Budin, L.: Dynamic scheduling with genetic programming. In: Collet, P., Tomassini, M., Ebner, M., Gustafson, S., Ekárt, A. (eds.) Genetic Programming, LNCS, vol. 3905, pp. 73–84. Springer, Berlin, Heidelberg (2006)
24. Jakobović, D., Marasović, K.: Evolving priority scheduling heuristics with genetic programming. Appl. Soft Comput. **12**(9), 2781–2789 (2012)
25. Jayamohan, M.S., Rajendran, C.: New dispatching rules for shop scheduling: a step forward. Int. J. Prod. Res. **38**, 563–586 (2000)
26. Jayamohan, M.S., Rajendran, C.: Development and analysis of cost-based dispatching rules for job shop scheduling. Eur. J. Oper. Res. **157**(2), 307–321 (2004)
27. Jedrzejowicz, P., Ratajczak-Ropel, E.: Agent-based gene expression programming for solving the RCPSP/max problem. In: Kolehmainen, M., Toivanen, P., Beliczynski, B. (eds.) Adaptive and Natural Computing Algorithms. Lecture Notes in Computer Science, vol. 5495, pp. 203–212. Springer, Berlin, Heidelberg (2009)
28. Johnston, M., Liddle, T., Zhang, M.: A relaxed approach to simplification in genetic programming. In: Esparcia-Alcázar, A.I., Ekárt, A., Silva, S., Dignum, S., Şima Uyar, A. (eds.) Genetic Programming, LNCS, vol. 6021, pp. 110–121. Springer, Berlin, Heidelberg (2010)
29. Koza, J.R.: Genetic Programming: On the Programming of Computers by Means of Natural Selection. MIT Press, Cambridge, MA (1992)
30. Kreipl, S.: A large step random walk for minimizing total weighted tardiness in a job shop. J. Sched. **3**, 125–138 (2000)
31. Kuczapski, A.M., Micea, M.V., Maniu, L.A., Cretu, V.I.: Efficient generation of near optimal initial populations to enhance genetic algorithms for job-shop scheduling. Inf. Technol. Control **39**(1), 32–37 (2010)
32. van Laarhoven, P.J.M., Aarts, E.H.L., Lenstra, J.K.: Job shop scheduling by simulated annealing. Oper. Res. **40**(1), 113–125 (1992)
33. Lourenco, H.R.: Job-shop scheduling: computational study of local search and large-step optimization methods. Eur. J. Oper. Res. **83**(2), 347–364 (1995)
34. McKay, K.N., Safayeni, F.R., Buzacott, J.A.: Job-shop scheduling theory: what is relevant? Interfaces **18**, 84–90 (1988)
35. Miyashita, K.: Job-shop scheduling with genetic programming. In: Whitley, D., Goldberg, D., Cantu-Paz, E., Spector, L., Parmee, I., Beyer, H.G. (eds.) In: GECCO 2000: Proceedings of the Genetic and Evolutionary Computation Conference, pp. 505–512. Morgan Kaufmann, San Francisco (2000)
36. Nguyen, S., Zhang, M., Johnston, M., Tan, K.: Learning iterative dispatching rules for job shop scheduling with genetic programming. Int. J. Adv. Manuf. Technol. **67**(1–4), 85–100 (2013)

37. Nguyen, S., Zhang, M., Johnston, M., Tan, K.C.: A computational study of representations in genetic programming to evolve dispatching rules for the job shop scheduling problem. IEEE Trans. Evol. Comput. **17**(5), 621–639 (2013)
38. Nguyen, S., Zhang, M., Johnston, M., Tan, K.C.: Dynamic multi-objective job shop scheduling: a genetic programming approach. In: Etaner-Uyar, A.Ş., Özcan, E., Urquhart, N. (eds.) Automated Scheduling and Planning, Studies in Computational Intelligence, vol. 505, pp. 251–282. Springer, Berlin, Heidelberg (2013)
39. Nguyen, S., Zhang, M., Johnston, M., Tan, K.C.: Learning reusable initial solutions for multi-objective order acceptance and scheduling problems with genetic programming. In: Krawiec, K., Moraglio, A., Hu, T., Etaner-Uyar, A.Ş., Hu, B. (eds.) Genetic Programming, LNCS, vol. 7831, pp. 157–168. Springer, Berlin, Heidelberg (2013)
40. Nguyen, S., Zhang, M., Johnston, M., Tan, K.C.: Automatic design of scheduling policies for dynamic multi-objective job shop scheduling via cooperative coevolution genetic programming. IEEE Trans. Evol. Comput. **18**(2), 193–208 (2014)
41. Nguyen, S.: Automatic design of dispatching rules for job shop scheduling with genetic programming. Ph.D. thesis, Victoria University of Wellington (2013)
42. Nie, L., Gao, L., Li, P., Li, X.: A GEP-based reactive scheduling policies constructing approach for dynamic flexible job shop scheduling problem with job release dates. J. Intell. Manuf. **24**(4), 763–774 (2013)
43. Nie, L., Shao, X., Gao, L., Li, W.: Evolving scheduling rules with gene expression programming for dynamic single-machine scheduling problems. Int. J. Adv. Manuf. Technol. **50**(5–8), 729–747 (2010)
44. Nie, L., Bai, Y., Wang, X., Liu, K.: Discover scheduling strategies with gene expression programming for dynamic flexible job shop scheduling problem. In: Tan, Y., Shi, Y., Ji, Z. (eds.) Adv. Swarm Intell. **7332**, 383–390 (2012)
45. Nowicki, E., Smutnicki, C.: A fast taboo search algorithm for the job shop problem. Manage. Sci. **42**, 797–813 (1996)
46. Ouelhadj, D., Petrovic, S.: A survey of dynamic scheduling in manufacturing systems. J. Sched. **12**(4), 417–431 (2009)
47. Petrovic, S., Fayad, C., Petrovic, D., Burke, E., Kendall, G.: Fuzzy job shop scheduling with lot-sizing. Ann. Oper. Res. **159**, 275–292 (2008)
48. Pickardt, C.W., Hildebrandt, T., Branke, J., Heger, J., Scholz-Reiter, B.: Evolutionary generation of dispatching rule sets for complex dynamic scheduling problems. Int. J. Prod. Econ. **145**(1), 67–77 (2013)
49. Pinedo, M.L.: Scheduling: Theory, Algorithms, and Systems, 3rd edn. Springer, New York (2008)
50. Pinedo, M., Singer, M.: A shifting bottleneck heuristic for minimizing the total weighted tardiness in a job shop. Naval Res. Logistics **46**(1), 1–17 (1999)
51. Ponnambalam, S.G., Ramkumar, V., Jawahar, N.: A multiobjective genetic algorithm for job shop scheduling. Prod. Plann. Control 12(8) (2001)
52. Potts, C.N., Strusevich, V.A.: Fifty years of scheduling: a survey of milestones. J. Oper. Res. Soc. **60**(Supplement 1), 41–68 (2009). http://www.palgrave-journals.com/jors/journal/v60/ns1/abs/jors20092a.html
53. Sels, V., Gheysen, N., Vanhoucke, M.: A comparison of priority rules for the job shop scheduling problem under different flow time- and tardiness-related objective functions. Int. J. Prod. Res. **50**(15), 4255–4270 (2011)
54. Sha, D., Hsu, C.Y.: A hybrid particle swarm optimization for job shop scheduling problem. Comput. Ind. Eng. **51**(4), 791–808 (2006)
55. Tay, J.C., Ho, N.B.: Evolving dispatching rules using genetic programming for solving multi-objective flexible job-shop problems. Comput. Ind. Eng. **54**(3), 453–473 (2008)
56. Wong, P., Zhang, M.: Algebraic simplification of gp programs during evolution. In: Proceedings of the 8th Annual Conference on Genetic and Evolutionary Computation. pp. 927–934. GECCO'06 (2006)

57. Xing, L.N., Chen, Y.W., Wang, P., Zhao, Q.S., Xiong, J.: A knowledge-based ant colony optimization for flexible job shop scheduling problems. Appl. Soft Comput. **10**(3), 888–896 (2010)

58. Yamada, T., Nakano, R.: A genetic algorithm with multi-step crossover for job-shop scheduling problems. In: GALESIA: First International Conference on Genetic Algorithms in Engineering Systems: Innovations and Applications. pp. 146–151 (1995)

59. Yin, W.J., Liu, M., Wu, C.: Learning single-machine scheduling heuristics subject to machine breakdowns with genetic programming. In: Sarker, R., Reynolds, R., Abbass, H., Tan, K.C., McKay, B., Essam, D., Gedeon, T. (eds.) In: The 2003 Congress on Evolutionary Computation (CEC 2003), vol. 2, pp. 1050–1055. IEEE Press, Piscataway, NJ (2003)

Evolutionary Fuzzy Systems: A Case Study for Intrusion Detection Systems

S. Elhag, A. Fernández, S. Alshomrani and F. Herrera

Abstract The so-called Evolutionary Fuzzy Systems consists of the application of evolutionary algorithms in the design process of fuzzy systems. Thanks to this hybridization, excellent abilities are provided to fuzzy systems in different work scenarios of data mining, such as standard classification, regression problems and association rule mining. The main reason of their success is the adaptation of their inner characteristics to any context. Among different areas of application, Evolutionary Fuzzy Systems have recently excelled in the area of Intrusion Detection Systems, yielding both accurate and interpretable models. To fully understand the nature and goodness of these type of models, we will introduce a full taxonomy on Evolutionary Fuzzy Systems. Then, we will overview a number of proposals from this research area that have been developed to address Intrusion Detection Systems. Finally, we will present a case study highlighting the good behaviour of Evolutionary Fuzzy Systems in this particular context.

Keywords Computational intelligence · Evolutionary fuzzy systems · Intrusion detection systems · Multi-objective evolutionary fuzzy systems · Fuzzy rule based systems

S. Elhag (✉)
Faculty of Computing and Information Technology,
King Abdulaziz University, Jeddah 21589, Saudi Arabia
e-mail: salma53ster@gmail.com

A. Fernández · F. Herrera
Department of Computer Science and Artificial Intelligence,
University of Granada, 18071 Granada, Spain
e-mail: alberto@decsai.ugr.es

F. Herrera
e-mail: herrera@decsai.ugr.es

S. Alshomrani
Faculty of Computing and Information Technology,
University of Jeddah, Jeddah 21589, Saudi Arabia
e-mail: sshomrani@kau.edu.sa

© Springer International Publishing AG, part of Springer Nature 2019
J. C. Bansal et al. (eds.), *Evolutionary and Swarm Intelligence
Algorithms*, Studies in Computational Intelligence 779,
https://doi.org/10.1007/978-3-319-91341-4_9

1   Introduction

As discussed in Chapter "Swarm and Evolutionary Computation" solutions based on Computational Intelligence [55] have shown a very high quality when applied to different problems on engineering, business, medicine and so on. Furthermore, when these techniques are used in synergy, i.e. combining their different components into a single robust model, the results are highly enhanced than when applying them in isolation. This fact has attracted the interest of many researchers on the topic. In particular, one of the most popular hybridizations is possibly the one between Fuzzy Rule Based Systems (FRBSs) [83] and Evolutionary Computation [41, 48] leading to Evolutionary Fuzzy Systems (EFSs) [18, 33].

The reason for the high success of EFS in solving problems, is that the learning procedure to determine the components of an FRBSs is usually carried out in an automated way. Therefore, this process is very likely to be addressed as an optimization problem, taking advantage of the capabilities of Evolutionary Algorithms (EAs) [24] as a robust global search technique. In addition to be very reliable techniques for complex problems, the generic code structure and independent performance features of EAs allow them to incorporate a priori knowledge. In the case of FRBSs, the former can be regarded from different perspectives, namely the definition of the fuzzy sets, the fuzzy membership function parameters, fuzzy rules, number of rules and many others. Furthermore, this approach has been extended by using Multi-Objective Evolutionary Algorithms (MOEAs) [16, 21], which can consider multiple conflicting objectives. The hybridization between MOEAs and FRBSs is currently known as Multi-Objective Evolutionary Fuzzy Systems (MOEFSs) [28].

As stated in the introduction of this work, there are many areas of application for Computational Intelligence and Soft Computing techniques. Among them, intrusion detection must be stressed as a very important task for providing security and integrity in information systems [85]. Analyzing the information gathered by security audit mechanisms, Intrusion Detection Systems (IDS) apply several rules that discriminate between legitimate events or an undesirable use of the system [3, 78].

In this area of research, fuzzy systems have shown to be a very valuable tool [8, 25, 62]. The reason is two-fold: first, the intrusion detection problem involves many numeric attributes, and models which are directly built on numeric data might cause high detection errors. Hence, small deviations in an intrusion might not be detected and small changes in the normal user profile may cause false alarms. Second, security itself includes fuzziness, as the boundary between the normal and abnormal behavior cannot be well defined.

However, in the context of IDS there are several metrics of performance to be optimized. Among others, we must stress the attack detection rate (ADR), which stands for the accuracy obtained for the attack classes managed as a whole, and the false alarm rate (FAR), i.e. the number of false positives. For the aforementioned reasons, the use of MOEFSs is a very well-suited approach to fulfill all the requirements needed to achieve a robust IDS [33, 60].

In this chapter, our first goal is to provide a clear definition of EFS, focusing on their main properties and presenting a complete taxonomy that comprise the main types of EFSs proposed in the specialized literature. Then, we focus on presenting the use of EFSs in IDS, providing a list of the most relevant contributions in this area of work. Finally, we will show the goodness of this type of approaches presenting a case study on the topic over a well-known IDS benchmarking problem using EFS and MOEFS algorithms [25, 26].

For achieving these objectives, the remainder of this chapter is organized as follows. In Sect. 2, we focus our attention to EFSs, presenting a complete taxonomy and providing examples of the different types. Section 3 is devoted to the application of EFSs in IDS, introducing the features of this problem, and enumerating those EFS approaches that have been designed for addressing this task. Next, in Sect. 4, we show a brief case study to excel the good behaviour of EFSs in this area. Finally, in Sect. 5, we provide some concluding remarks of this work as well as providing several challenges for future work on the topic of EFS.

2 Evolutionary Fuzzy Systems: Taxonomy and Analysis

The essential part of FRBSs is a set of IF-THEN fuzzy rules (traditionally linguistic values), whose antecedents and consequents are composed of fuzzy statements, related to with the dual concepts of fuzzy implication and the compositional rule of inference. Specifically, an FRBS is composed of a *knowledge base* (KB), that includes the information in the form of those IF-THEN fuzzy rules, i.e. the Rule Base (RB), and the correspondence of the fuzzy values, known as Data Base (DB). It also comprises of an inference engine module that includes a fuzzification interface, an *inference system*, and a defuzzification interface.

EFSs is a family of approaches that are built on top of FRBSs, whose components are improved by means of an evolutionary learning/optimization process as depicted in Fig. 1. This process is designed for acting or tuning the elements of a fuzzy system in order to improve its behavior in a particular context. Traditionally, this was carried out by means of Genetic Algorithms, leading to the classical term of Genetic Fuzzy Systems [17, 18, 20, 45]. In this chapter, we consider a generalization of the former by the use of EAs [24].

The central aspect on the use of EAs for automatic learning of FRBSs is that the design process can be analyzed as a search problem in the space of models, such as the space of rule sets, membership functions, and so on. This is carried out by means of the coding of the model in a chromosome. Therefore, the first step in designing an EFS is to decide which parts of the fuzzy system are subject to optimization by the EA coding scheme. Hence, EFS approaches can be mainly divided into two types of processes: tuning and learning. Additionally, we must make a decision whether to just improve the accuracy/precision of the FRBS or to achieve a tradeoff between accuracy and interpretability (and/or other possible objectives) by means of a MOEA. Finally, we must stress that new fuzzy set representations have been designed, which

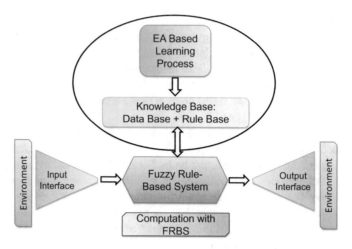

Fig. 1 Integration of an EFS on top of an FRBS

implies a new aspect to be evolved in order to take the highest advantage of this approach.

This high potential of EFSs implies the development of many different types of approaches. In accordance with the above, and considering the FRBSs' components involved in the evolutionary learning process, a taxonomy for EFS was proposed by Herrera in [45] (please refer to its thematic Website at http://sci2s.ugr.es/gfs/). More recently, in [33] authors extended the former by distinguishing among the learning of the FRBSs' elements, the EA components and tuning, and the management of the new fuzzy sets representation. This novel EFS taxonomy is depicted in Fig. 2.

In order to describe this taxonomy tree of EFSs, this section is arranged as follows. First, we present these models according to the FRBS components involved in the evolutionary learning process (Sect. 2.1). Afterwards, we focus on the multi-objective optimization (Sect. 2.2). Finally, we provide some brief remarks regarding the parametrized construction for new fuzzy representations (Sect. 2.3).

2.1 Evolutionary Learning and Tuning of FRBSs' Components

When addressing a given Data Mining problem, the use of any fuzzy sets approach is usually considered when certain requirements are pursued. First, when an interpretable system is sought; second, when the uncertainty involved in the data must be properly managed; finally, even when a dynamic model is under consideration. Then, we must make the decision on whether a simple FRBS is enough for the

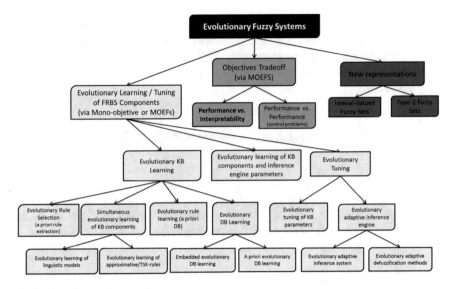

Fig. 2 Evolutionary fuzzy systems taxonomy

given requirements, or if a more sophisticated solution is needed, thus exchanging computational time for accuracy.

This can be achieved either by two different ways. On the one hand, by designing approaches to learn the KB components, including an adaptive inference engine. On the other hand, by starting from a given FRBS, developing approaches to tune the aforementioned components. Therefore, we may distinguish among the evolutionary KB learning, the evolutionary learning of KB components and inference engine parameters, and the evolutionary tuning. These approaches are described below, which can be performed via a standard mono-objective approach or a MOEA.

2.1.1 Evolutionary KB Learning

The following four KB learning possibilities can be considered:

1. *Evolutionary rule selection.* In order to get rid of irrelevant, redundant, erroneous and/or conflictive rules in the RB, which perturb the FRBS performance, an optimized subset of fuzzy rules can be obtained [51].
2. *Simultaneous evolutionary learning of KB components.* Working in this way, there is possibility of generating better definitions of these components [49]. However, a larger search space is associated with this case, which makes the learning process more difficult and slow.
3. *Evolutionary rule learning.* Most of the approaches proposed to automatically learn the KB from numerical information have focused on the RB learning, using a predefined DB [75].

4. *Evolutionary DB learning.* A DB generation process allows the shape or the membership functions to be learnt, as well as other DB components such as the scaling functions, the granularity of the fuzzy partitions, and so on. Two possibilities can be used: "a priori evolutionary DB learning" and "embedded evolutionary DB learning [19]."

2.1.2 Evolutionary Learning of KB Components and Inference Engine Parameters

This area belongs to a hybrid model between adaptive inference engine and KB components learning. These type of approaches try to find high cooperation between the inference engine via parameters adaptation and the learning of KB components, including both in a simultaneous learning process [59].

2.1.3 Evolutionary Tuning

With the aim of making the FRBS perform better, some approaches try to improve the preliminary DB definition or the inference engine parameters once the RB has been derived. The following three tuning possibilities can be considered (see the sub-tree under "evolutionary tuning").

1. *Evolutionary tuning of KB parameters.* A tuning process considering the whole KB obtained is used a posteriori to adjust the membership function parameters, i.e. the shapes of the linguistic terms [11].
2. *Evolutionary adaptive inference systems.* This approach uses parameterized expressions in the inference system, sometimes called adaptive inference systems, for getting higher cooperation among the fuzzy rules without losing the linguistic rule interpretability [6].
3. *Evolutionary adaptive defuzzification methods.* When the defuzzification function is applied by means of a weighted average operator, i.e. parameter based average functions, the use of EAs can allow us to adapt these defuzzification methods [54].

2.2 *Approaches for Optimizing Several Objectives*

Traditionally, the efforts in developing EFSs were aimed at improving the accuracy/precision of the FRBS in a mono-objective way. However, in current applications the interest of researchers in obtaining more interpretable linguistic models has significantly grown [39]. The hitch is that accuracy and interpretability represent contradictory objectives. A compromise solution is to address this problem using MOEAs [16] leading to a set of fuzzy models with different tradeoffs between both objectives instead of a biased one. These hybrid approaches are known as MOEFSs [28] that, in addition to the two aforementioned goals, may include any other kind of objective, such as the complexity of the system, the cost, the computational time, additional performance metrics, and so on [61].

In this case, the division of these type of techniques is first based on the multi-objective nature of the problem faced and second on the type of FRBS components optimized. Regarding the previous fact, those of the second level present a clear correspondence with the types previously described for EFSs in the previous section.

Here, we will only present a brief description for each category under considera-tion. For more detailed descriptions or an exhaustive list of contributions see [28] or its associated Webpage (http://sci2s.ugr.es/moefs-review/).

2.2.1 Accuracy-Interpretability Trade-Offs

The comprehensibility of fuzzy models began to be integrated into the optimization process in the mid 1990s [50], thanks to the application of MOEAs to fuzzy systems. Nowadays, researchers agree on the need to consider two groups of interpretability measures, complexity-based and semantic-based ones. While the first group is related to the dimensionality of the system (simpler is better) the second one is related to the comprehensibility of the system (improving the semantics of the FRBS components) [68]. Some recent applications show the significance of balancing both the ability to adequately represent the decision making processes with the ability to provide a domain user with compact and understandable explanation and justification of the proposed decisions [43].

The differences between both accuracy and interpretability influence the optimiza-tion process, so researchers usually include particular developments in the proposed MOEA making it able to handle this particular trade-off. An example can be seen in [38] where authors specifically force the search to focus on the most accurate solutions. For a complete survey on interpretability measures for linguistic FRBSs see [39].

2.2.2 Performance Versus Performance (Control Problems)

In control system design, there are often multiple objectives to be considered, i.e. time constraints, robustness and stability requirements, comprehensibility, and the compactness of the obtained controller. This fact has led to the application of MOEAs in the design of Fuzzy Logic Controllers.

The design of these systems is defined as the obtaining of a structure for the controller and the corresponding numerical parameters. In a general sense, they fit with the tuning and learning presented for EFSs in the previous section. In most cases, the proposal deals with the postprocessing of Fuzzy Logic Controller parameters, since it is the simplest approach and requires a reduced search space.

2.3 Novel Fuzzy Representations

Classical approaches on FRBSs make use of standard fuzzy sets [84], but in the specialized literature we found extensions to this approach with aim to better represent the uncertainty inherent to fuzzy logic. Among them, we stress Type-2 fuzzy sets [52] and Interval-Valued Fuzzy Sets (IVFSs) [67] as two of the main exponents of new fuzzy representations.

Type-2 fuzzy sets reduce the amount of uncertainty in a system because this logic offers better capabilities to handle linguistic uncertainties by modeling vagueness and unreliability of information. In order to obtain a type-2 membership function, we start from the type-1 standard definition, and then we blur it to the left and to the right. In this case, for a specific value, the membership function, takes on different values, which are not all weighted the same. Therefore, we can assign membership grades to all of those points.

For IVFS [67], the membership degree of each element to the set is given by a closed sub-interval of the interval [0, 1]. In such a way, this amplitude will represent the lack of knowledge of the expert for giving an exact numerical value for the membership. We must point out that IVFSs are a particular case of type-2 fuzzy sets, having a zero membership out of the ranges of the interval.

In neither case, there is a general design strategy for finding the optimal fuzzy models. In accordance with the former, EAs have been used to find the appropriate parameter values and structure of these fuzzy systems.

In the case of type-2 fuzzy models, EFSs can be classified into two categories [12]: (1) the first category assumes that an "optimal" type-1 fuzzy model has already been designed, and afterwards a type-2 fuzzy model is constructed through some sound augmentation of the existing model [13]; (2) the second class of design methods is concerned with the construction of the type-2 fuzzy model directly from experimental data [58].

Regarding IVFS, current works initialize type-1 fuzzy sets as those defined homogeneously over the input space. Then, the upper and lower bounds of the interval for each fuzzy set are learned by means of a weak-ignorance function (amplitude tuning) [69], which may also involve a lateral adjustment for the better contextualization of the fuzzy variables [71]. Finally, in [70] IVFS are built ad-hoc, using an interval-valued restricted equivalence functions within a new interval-valued fuzzy reasoning method. The parameters of these equivalence functions per variable are learned by means of an EA, which is also combined with rule selection in order to decrease the complexity of the system.

3 The Use of Evolutionary Fuzzy Systems for Intrusion Detection Systems

In different application areas where a given model must automatically built with respect to the available data, the requirements tend to be quite similar. First, the output model should be interpretable by the final user, i.e. it should easily allow to explain the phenomena that has been identified. Second, it must have the ability to adapt properly to different optimization strategies. Finally, it should be able to extract the hidden knowledge with a good trade-off between recall and precision.

For these reasons, the use of EFSs is so much extended in a wide number of scenarios. Among them, IDS has gained a major interest due to the rise of on-line services and communications, and the need of providing security and integrity in these information systems.

In this section, we will first present a summary of the main concepts for IDS (Sect. 3.1). Then, we will overview some of the approaches that have been developed to address this problem with EFS (Sect. 3.2).

3.1 Background on Intrusion Detection Systems

In this data age we are witnessing how computer systems are creating, processing, and sharing an overwhelming quantity of information. According to this fact, computer security must be regarded as a critical issue, so that the unauthorized access to this data from a computer and/or computer network, could imply a significant problem, as it compromises the integrity, confidentiality and availability of the resources [14]. This issue is of extreme importance in recent application areas due to the digitalization and the Internet of Things [85]. Therefore, a wide amount of computer security tools such as antiviruses, firewalls, data encryption, have been introduced. In addition to this, there are some complementary tools that monitor the activity of the network in order to detect and block intrusions.

Anomalous activities are thus identified by IDSs, which comprise the process of monitoring and analyzing events occurring in a computer system or network in order to detect anomalous activity [78].

IDS can be split into two categories according to the detection methods they employ, including (1) misuse detection and (2) anomaly detection. The main difference between both types of systems is related to whether they use a signature detection or anomaly detection paradigm. Misuse detection systems take the majority of IDSs, and use an established set of known attack patterns, and then monitor the net trying to match incoming packets and/or command sequences to the signatures of known attacks [57]. Hence, decisions are made based on the prior knowledge acquired from the model. The main advantage of this type of IDS is that they provide high detection accuracy with few false positives, but with the disadvantage that they are not able to detect new attacks other than those previously stored in the database.

On the other hand, anomaly detection IDS have the ability to detect new attacks, but at the cost of increasing the number of false positives. In an initial phase, the anomaly based IDS is trained in order to obtain a normal profile of activity in the system [64]. The learned profiles of normal activity are customized for every system, making it quite difficult for an attacker to know with certainty what activities it can carry out without getting detected. Then, incoming traffic is processed in order to detect variations in comparison with the normal activity, in which case it will be considered as a suspicious activity. In addition to the higher number of false alarms raised, another disadvantage of the development a system of these characteristics is the higher the complexity compared to the case of misuse detection.

3.2 Related Work for Fuzzy Systems in IDS

The ultimate goal of IDS is to achieve a high attack detection rate along with a low false alarm rate, being this a serious challenge to be overcome. For this reason, both misuse detection and anomaly detection system make use of Data Mining techniques to aid in the processing of large volumes of audit data and the increasing complexity of intrusion behaviors [65, 86].

In particular, Soft Computing and Computational Intelligence techniques have become essential pieces for addressing this problem [82]. Among all techniques in this paradigm, the properties of fuzzy logic for the development of IDS must be taken into consideration. As stated in the introduction of this work, the reason is two-fold. On the one hand, if we focus on the feature space of IDS applications, we may observe that it usually comprises many different variables. Therefore, the use of linguistic variables allows at condensing the information as well as representing uncertain knowledge associated to IDS. On the other hand, the output space includes several types of categories, i.e. a large number of attack events. The smoothness associated with fuzzy logic can provide more confident rules in the areas of overlapping.

For the aforementioned reasons, throughout the years many approaches have been proposed and analyzed aiming to take advantage of these fuzzy systems. One of the first techniques was the Fuzzy Intrusion Recognition Engine (FIRE) [22, 23]. This approach employ the well known C-means algorithm for defining the fuzzy sets and their membership functions, and then authors determine their own hand-encoded rules for malicious network activities, which was probably the main limitation of this work.

Regarding EFS, to the best of our knowledge few works have been published in the specialized literature that address this area. For example, in [42] a genetic programming algorithm evolves tree-like structure of chromosomes (rules) whose antecedents are composed of triangular membership functions. Multiple objective functions are defined, which are then combined into a single fitness function by means of user-defined weights. The hitch here is that these weights cannot be optimized dynamically for different cases.

A deep study of different architectures for EFS have been developed in [1, 2]. In these works, fuzzy rules are expressed in the generic way: antecedent of fuzzy labels, consequent with class and rule weight. Then, authors analyze the three main schemes for rule generation with genetic' algorithms, namely the Genetic Cooperative-Competitive Learning (GCCL) [44], the Pittsburgh approach [72, 73], and the Iterative Rule Learning (IRL) [79]. Additionally, in [80] authors extended the previous work by defining a parallel environment for the execution of the population of rules.

Another topic of work is the integration of association rules and frequent episodes with fuzzy logic [37]. In one of the latest publications [74], authors use Apriori as baseline algorithm and fuzzify the obtained rules following the recommendations made in [56]. Then, several implementation techniques were used to speed up the algorithm, i.e. to reduce items involved in rule induction without resulting into any considerable information loss.

The interest on the use of EFS have been also shown in the field of fuzzy association mining [63]. In this latter work, the procedure is divided into two stages: (1) authors generate a large number of candidate association fuzzy rules for each class; (2) with aims at reducing the fuzzy rule search space, a boosting GA based on the IRL approach is applied for each class for rule pre-screening using two evaluation criteria. However, it only optimizes classification accuracy and omits the necessity of interpretability optimization.

A recent work on this topic [26], focused on the synergy of a robust fuzzy associative classifier (FARC-HD) [5] with the One-vs-One (OVO) class decomposition technique [40], in which binary subproblems are obtained by confronting all possible pair of classes. The high potential of this fuzzy rule learning approach was determined by the goodness in the correct identification for all types of attacks, including rare attack categories.

Finally, MOEFS have also been analyzed in the context of IDS. In [77] the authors propose MOGFIDS (short for Multi-Objective Genetic Fuzzy Intrusion Detection System), which is based on the previous work of the authors related to an agents-based evolutionary approach for fuzzy rules [81]. This approach is based on the construction and evolution, in a Pittsburgh style, of an accurate and interpretable fuzzy knowledge base. Specifically, it is a genetic wrapper that searches for a near-optimal feature subset from network traffic data.

One of the latest proposals in this field was use of an MOEFS in which the genetic optimization was focused on carrying out a rule selection and DB tuning [25]. The aim for this procedure was to be able at both extending the search space and obtaining a wide amount of accurate solutions. By doing so, the final user may select the most suitable classification system for the current work context.

4 Case Study: Addressing Intrusion Detection Systems with Multi-objective Evolutionary Fuzzy Systems

In this section we aim to show the goodness of MOEFSs to address the problem of IDS. Specifically, we will present how this type of system is capable of reaching a high global performance under different metrics of interest for IDS application, as well as providing a simple and compact knowledge model. To do so, we will carry out a brief experimental study with the well-known KDDCUP'99 dataset, whose features are presented in Sect. 4.1. As EFSs, we have selected the FARC-HD classifier [5] and its extensions to IDS proposed in [25, 26], whose configuration is given in Sect. 4.2. The metrics of performance considered to analyze the behavior of these models are presented in Sect. 4.3. Finally, the experimental results are shown in Sect. 4.4.

4.1 Benchmark Data: KDDCUP'99 Problem

Among different benchmark problems for IDS, the KDDCUP'99 dataset is possibly the most used one, being a standard until today [9, 15, 53]. It was obtained by the Information System Technology (IST) group of Lincoln laboratories at MIT University under contract of DARPA and in collaboration with ARFL [57]. It consisted of an environment of a local area network (LAN) that simulates a typical U.S. Air Force LAN, including several weeks of raw TCP dump data with normal activities and various types of attacks.

It comprises 41 attributes in total, which are divided three main groups: intrinsic features (extracted from the headers' area of the network packets), content features (extracted from the contents area of the network packets), traffic features (extracted with information about previous connections).

Class labels are divided into normal and attack activities. This last class can be further divided into particular types of attack, which are basically grouped into four major categories, namely:

- Denial of Service (DOS): make some machine resources unavailable or too busy to answer to legitimate users requests (SYN flooding).
- Probing (PRB): Surveillance for information gathering or known vulnerabilities about a network or a system (port scanning).
- Remote To Local (R2L): use vulnerability in order to obtain unauthorized access from a remote machine (password guessing).
- User To Root (U2R): exploit vulnerabilities on a system to gain local super-user (root) privileges (buffer overflow attack).

In this dataset, the total amount of data places it in the context of Big Data [34], i.e. affecting the scalability of current approaches. For this reason, usually a small portion of the whole data is randomly selected for its use with standard classifiers. Specifically, we will select just a 10% of the instances for our experiments. This

Table 1 Number of examples per class in each dataset partition for KDDCUP'99 problem

Class	KDDCUP'99	
	#Ex. training	#Ex. test
Normal	8783	79,049
DOS	5457	49,115
PRB	213	1917
R2L	100	899
U2R	26	26
Total	14,579	131,006

implies a total of 494,021 connections. Then, we have also removed all duplicate instances, reducing the data to a total of 145,585 examples.

Finally, in order to carry out a validation procedure of the results, we have selected a hold-out methodology. Specifically, we will employ a 10% of the datasets for training and the remaining 90% for test. However, in order to take into account the original distribution of classes, we will include a 50% of instances for U2R in both training and test. Table 1 shows the final distribution of examples for each partition/class.

4.2 Algorithms and Parameters

As stated in the begging of this section, we have considered several EFS algorithms that have shown a good behavior for IDS problems. Specifically, all of them are based on the standard FARC-HD classifier [5]. The first one, is a multi-classifier extension, named as FARC-HD-OVO [26]. The second one is a MOEFS noted as FARC-HD-MOEA [25], include a NSGA-II optimization procedure for the tuning of the KB according to different IDS metrics. Additionally, we will include C4.5 [66] in the experimental study as a state-of-the-art rule induction algorithm. In what follows, we detail the configuration of the parameters for each approach:

1. **FARC-HD** [5]: First, we have selected 5 labels per variable for the fuzzy sets, product t-norm as conjunction operator and additive combination for the inference procedure. As specific parameters of the learning stage, we have set up the minimum support to 0.05 and the minimum confidence to 0.8. Finally, we have fixed the maximum depth of the tree to a value of 3, and the k parameter for the pre-screening to 2. For more details about these parameters, please refer to [5].
   We must stress that this configuration will be shared for all three models based on FARC-HD, i.e. the standard approach, FARC-HD-OVO, and FARC-HD-MOEA.
2. **FARC-HD-OVO** [7]: The learning procedure will be performed using all possible pairs of classes. In order to aggregate the outputs of each binary classifier into

a single solution, we will make use of the preference relations solved by Non-Dominance Criterion (ND) [35].

3. **FARC-HD-MOEA**: The parameters of the NSGA-II MOEA have been set up as follows: 50 individuals as population size, with 20,000 generations. The crossover and the mutation (per gen) probabilities are 0.9 and 0.025 respectively. The objectives/metrics selected for the tuning are those that shown the best behavior in [25], namely MfM and FAR.

4. **C4.5** [66]: For C4.5 we have set a confidence level of 0.25, the minimum number of item-sets per leaf was set to 2 and the application of pruning was used to obtain the final tree. We must point out that, for the sake of allowing the output model to be compact and interpretable, we have carried out an extensive pruning. Specifically, we have limited the maximum depth of the tree to 3. Therefore, rules obtained from C4.5 will be of the same length than those learned by the FARC-HD algorithms, establishing a fair comparison between both techniques.

4.3 Performance Metrics for IDS

In the specialized literature for IDS in general, and for misuse detection in particular, authors have made use of several metrics of performance for the evaluation of their results in comparison with the state-of-the-art. In this chapter, we have selected different measures which will allow us to analyze the behaviour of our approach under several perspectives:

1. *Accuracy*: It stands for the global percentage of hits. In our case (IDS), its contribution is low as it does not take into account the individual accuracies of each class, but it has been selected as a classical measure.

$$Acc = \frac{\sum_{i=1}^{C} TP_i}{N} \tag{1}$$

where C is the number of classes, N is the number of examples and TP_i is the number of True Positives of the i-th class.

2. *Mean F-Measure*. In the binary case, the standard f-measure computes a trade-off between precision and recall of both classes. In this case, we compute the average for the F-measure achieved for each class (taken as positive) and the remaining ones (taken as a whole as negative):

$$MFM = \frac{\sum_{i=1}^{C} FM_i}{C} \tag{2}$$

$$FM_i = \frac{2 \cdot Recall_i \cdot Precision_i}{Recall_i + Precision_i} \tag{3}$$

$$Precision_i = \frac{TP_i}{TP_i + FP_i} \qquad (4)$$

$$Recall_i = \frac{TP_i}{TP_i + FN_i} \qquad (5)$$

where TP_i, FP_i and FN_i are the number of true positives, false positives and false negatives of the i-th class respectively. percentage).

3. *Average accuracy.* It is computed as the average of the individual hits for each class. For this reason, it is also known as the average recall:

$$AvgAcc = \frac{1}{C} \sum_{i=1}^{C} Recall_i \qquad (6)$$

4. *Attack Accuracy.* In this case we omit the "Normal" instances and we focus in checking whether we guess correctly the different "Attack" types individually.

$$AttAcc = \frac{1}{C-1} \sum_{i=2}^{C} Recall_i \qquad (7)$$

In this case, the first class ($i = 1$) is considered to be the "Normal" class.

5. *Attack Detection Rate.* It stands for the accuracy rate for the attack classes. Therefore, it is computed as:

$$ADR = \frac{\sum_{i=2}^{C} TP_i}{\sum_{i=2}^{C} TP_i + FN_i} \qquad (8)$$

Reader must take into account that also in this case, the first class ($i = 1$) is considered to be the "Normal" class.

6. *False Alarm Rate.* In this case, we focus on the "Normal" examples, and we check which is the percentage of "false negatives" found, i.e. those instances identified as "alarms" but which are actually normal behavior.

$$FAR = \frac{FP_1}{TP_1 + FP_1} \qquad (9)$$

As in the former metric (*ADR*), the "Normal" class has the first index ($i = 1$).

4.4   Experimental Results

All performance values of interest on IDS that were obtained by the different classifiers in the KDDCUP'99 dataset are shown in Table 2. Best values for each metric is stressed in boldface.

Analyzing these results, we observe that FARC-HD-MOEA achieves a significant improvement over the standard FARC-HD method in most of the considered metrics of performance. We must recall that the same configuration is shared by both approaches. In other words, the initial KB is exactly the same, and it is the optimization procedure what truly excels the behavior of the novel approach, implying the goodness in the design and capabilities of the MOEA optimization procedure versus the standard Genetic Algorithm when dealing with IDS problems. In particular, we must stress the differences with respect to the values of the mean f-measure, average accuracy, and attack accuracy are especially remarkable, improving up to 10–15 points in some cases.

If we contrast the behavior of FARC-HD-MOEA versus the multi-classifier FARC-HD-OVO, the differences are reduced. The benefit of the FARC-HD-MOEA must be regarded in terms of the simplicity and interpretability in using a single classifier, instead of a whole ensemble. As stated, the advantage is two-fold. On the one hand, the efficiency in the system response during the inference. On the other hand, the expert is able to analyze the rule(s) associated with each corresponding decision.

When analyzing FARC-HD-MOEA versus the C4.5 decision tree, an interesting behavior is observed. Whereas global metrics of performance such as accuracy and/or attack detection rate are usually higher for C4.5, the goodness of FARC-HD-MOEA lies in the ability of providing a good average recognition. This issue is evident when observing the value of the mean f-measure and the average accuracy.

Analyzing the results from another perspective, we may determine that a low number of simple (compact) linguistic rules are enough to cover the whole problem

Table 2   Complete experimental results for the EFS classifiers (FARC-HD-MOEA, FARC-HD and FARC-HD-OVO), and C4.5 over the reduced KDDCUP'99 dataset for different metrics of performance: Accuracy (Acc), Mean F-measure (MFM), Average accuracy (AvgAcc), Attack average accuracy (AttAcc), Attack detection rate (ADR), and False alarm rate (FAR)

Metric	FARC-HD-MOEA		FARC-HD		FARC-HD-OVO		C4.5	
	Tr	Tst	Tr	Tst	Tr	Tst	Tr	Tst
Acc	98.11	97.89	98.42	98.30	99.18	99.00	99.49	**99.44**
MFM	91.99	**86.06**	90.69	84.26	97.72	84.12	92.96	80.85
AvgAcc	89.57	89.30	88.31	87.76	96.50	**89.32**	91.20	86.84
AttAcc	87.06	**86.77**	85.44	84.77	95.64	86.70	89.04	83.61
ADR	95.84	95.53	96.27	96.17	98.07	97.77	98.96	**98.93**
FAR	0.3871	0.5528	0.1708	0.2948	0.0797	**0.1910**	0.1594	0.2277

Table 3 Comparison of number of rules (#Rules) and average number of antecedents (#Avg. Ant.) for the algorithms selected in the experimental study

Dataset	FARC-HD-MOEA		FARC-HD		FARC-HD-OVO		C4.5	
	#Rules	#Avg. Ant.	#Rules	#Avg. Ant.	#Rules	#Avg. Ant.	#Rules	#Avg. Ant.
KDDCUP'99	44	2.6590	25	2.3600	84	2.2238	150	2.1385

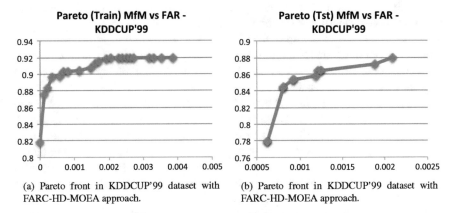

(a) Pareto front in KDDCUP'99 dataset with FARC-HD-MOEA approach.

(b) Pareto front in KDDCUP'99 dataset with FARC-HD-MOEA approach.

Fig. 3 Pareto front obtained in the test stage with FARC-HD-MOEA approach. Objectives selected during the search were the mean F-measure (MFM) and the false alarm rate (FAR)

space accurately. Specifically, in Table 3 we may observe the comparison in total number of rules and average number of antecedents in the case of the EFS algorithms and C4.5.

Finally, for the sake of complementing this study, we show in Fig. 3 the complete Pareto front obtained from the optimization procedure in the KDDCUP'99 dataset. We may observe a wide amount of non-dominated solutions from both the training and test sets, all of which are homogeneously distributed in the solution space. This issue reflects the good properties of the search procedure, as it covers a wide amount of different cases from which the expert can select the most appropriate one for a desired profile of behaviour.

5   Conclusions and Future Perspectives

In this chapter, we have reviewed the topic of EFSs focusing on the application of this type of systems for IDS. To have a clear picture on EFS, we have introduced a complete taxonomy for the current types of associated methodologies. Then, we have focused on the topic of IDS, identifying its main characteristics and providing some examples of solutions based on EFS that have been successful in this area. Finally, we have carried out a short experimental study with the well-known KDDCUP'99 dataset in order to contrast the behavior of three different EFS based on the FARC-HD algorithm, and the C4.5 decision tree. The obtained results using several metrics of performance in the scenario of IDS shown the goodness of EFS over C4.5, both in terms of accuracy and interpretability.

But in spite of the high performance shown by EFS in this context, we must acknowledge that there is still room for improvement in this paradigm of models, especially regarding new areas of application. For example, we must be aware of the novel non-standard and complex classification problems that have gathered a significant attention in the specialized literature. We are referring to ordinal and monotonic classification [10], multi-instance [47], and multi-label learning [46]. At present, just few works using EFS have been proposed [4], implying a clear gap with respect to standard approaches.

At present, one of the hottest topics for research is related to Data Science and Big Data problems [31]. An in depth analysis of the current state of this framework was carried out in both [30, 32], where authors investigate the good properties of fuzzy systems when devoted to solve such applications. However, focusing on the case of EFS for Big Data, the evolutionary procedure related to its core implies a constraint for the development of scalable solutions. Therefore, also few works are yet developed in this area of research [29, 36].

Finally, the optimization of the inner components of FRBS must be still investi-gated to develop better models. Some very interesting recent works have focused on the aggregation operations [27]. In addition to the definition of the fuzzy system, one should also focus on the elements of the EAs, namely the use of novel techniques [76] or extension of standard GA components such as niching GAs for multimodal func-tions, among others. However, must stress that a justification for their choice must be made from whatever meaningful point of view: efficiency, efficacy/precision, interpretability, scalability, and so on.

References

1. Abadeh, M.S., Mohamadi, H., Habibi, J.: Design and analysis of genetic fuzzy systems for intrusion detection in computer networks. Expert Syst. Appl. **38**(6), 7067–7075 (2011)
2. Abadeh, M.S., Habibi, J., Lucas, C.: Intrusion detection using a fuzzy genetics-based learning algorithm. J. Netw. Comput. Appl. **30**(1), 414–428 (2007)
3. Aburomman, A., Reaz, M.: A survey of intrusion detection systems based on ensemble and hybrid classifiers. Comput. Secur. **65**, 135–152 (2017)
4. Alcala-Fdez, J., Alcala, R., Gonzalez, S., Nojima, Y., Garcia, S.: Evolutionary fuzzy rule-based methods for monotonic classification. IEEE Trans. Fuzzy Syst. **25**(6), 1376–1390 (2017)
5. Alcala-Fdez, J., Alcala, R., Herrera, F.: A fuzzy association rule-based classification model for high-dimensional problems with genetic rule selection and lateral tuning. IEEE Trans. Fuzzy Syst. **19**(5), 857–872 (2011)
6. Alcala-Fdez, J., Herrera, F., Marquez, F.A., Peregrin, A.: Increasing fuzzy rules cooperation based on evolutionary adaptive inference systems. International Journal of Intelligent Systems **22**(9), 1035–1064 (2007)

7. Alshomrani, S., Bawakid, A., Shim, S.O., Fernandez, A., Herrera, F.: A proposal for evolutionary fuzzy systems using feature weighting: dealing with overlapping in imbalanced datasets. Knowl. -Based Syst. **73**, 1–17 (2015)
8. Ashfaq, R., Wang, X.Z., Huang, J., Abbas, H., He, Y.L.: Fuzziness based semi-supervised learning approach for intrusion detection system. Inf. Sci. **378**, 484–497 (2017)
9. Benferhat, S., Boudjelida, A., Tabia, K., Drias, H.: An intrusion detection and alert correlation approach based on revising probabilistic classifiers using expert knowledge. Appl. Intell. **38**(4), 520–540 (2013)
10. Cardoso, J.S., Sousa, R.: Measuring the performance of ordinal classification. Int. J. Pattern Recogn. Artif. Intell. **25**(8), 1173–1195 (2011)
11. Casillas, J., Cordon, O., del Jesus, M.J., Herrera, F.: Genetic tuning of fuzzy rule deep structures preserving interpretability and its interaction with fuzzy rule set reduction. IEEE Trans. Fuzzy Syst. **13**(1), 13–29 (2005)
12. Castillo, O., Melin, P.: Optimization of type-2 fuzzy systems based on bio-inspired methods: a concise review. Inf. Sci. **205**, 1–19 (2012)
13. Castillo, O., Melin, P., Garza, A.A., Montiel, O., Sepulveda, R.: Optimization of interval type-2 fuzzy logic controllers using evolutionary algorithms. Soft Comput. **15**(6), 1145–1160 (2011)
14. Chebrolu, S., Abraham, A., Thomas, J.P.: Feature deduction and ensemble design of intrusion detection systems. Comput. Secur. **24**(4), 295–307 (2005)
15. Chung, Y.Y., Wahid, N.: A hybrid network intrusion detection system using simplified swarm optimization (SSO). Appl. Soft Comput. **12**(9), 3014–3022 (2012)
16. Coello-Coello, C.A., Lamont, G., van Veldhuizen, D.: Evolutionary Algorithms for Solving Multi-objective Problems, Genetic and Evolutionary Computation, 2nd edn. Springer, Berlin, Heidelberg (2007)
17. Cordon, O., Gomide, F., Herrera, F., Hoffmann, F., Magdalena, L.: Ten years of genetic fuzzy systems: current framework and new trends. Fuzzy Sets Syst. **141**, 5–31 (2004)
18. Cordon, O., Herrera, F., Hoffmann, F., Magdalena, L.: Genetic fuzzy systems. In: Evolutionary Tuning and Learning of Fuzzy Knowledge Bases. World Scientific, Singapore, Republic of Singapore (2001)
19. Cordon, O., Herrera, F., Villar, P.: Generating the knowledge base of a fuzzy rule-based system by the genetic learning of data base. IEEE Trans. Fuzzy Syst. **9**(4), 667–674 (2001)
20. Cordon, O.: A historical review of evolutionary learning methods for mamdani-type fuzzy rule-based systems: designing interpretable genetic fuzzy systems. Int. J. Approx. Reasoning **52**(6), 894–913 (2011)
21. Deb, K.: Multi-objective Optimization Using Evolutionary Algorithms. Wiley, Chichester, New York (2001)
22. Dickerson, J., Dickerson, J.: Fuzzy network profiling for intrusion detection. In: Proceedings of the 19th International Conference of the North American Fuzzy Information Society (NAFIPS'00). pp. 301–306. IEEE Press, Atlanta, GA, USA (2000)
23. Dickerson, J., Juslin, J., Koukousoula, O., Dickerson, J.: Fuzzy intrusion detection. In: Proceedings of the 20th International Conference of the North American Fuzzy Information Society (NAFIPS'01) and Joint the 9th IFSA World Congress. vol. 3, pp. 1506–1510. IEEE Press, Vancouver, Canada (2001)
24. Eiben, A.E., Smith, J.E.: Introduction to Evolutionary Computation. Springer, Berlin, Germany (2003)
25. Elhag, S., Fernández, A., Altalhi, A., Alshomrani, S., Herrera, F.: On the combination of genetic fuzzy systems and pairwise learning for improving detection rates on intrusion detection systems. Soft Comput. 1–16 (2018) (in press)
26. Elhag, S., Fernández, A., Bawakid, A., Alshomrani, S., Herrera, F.: On the combination of genetic fuzzy systems and pairwise learning for improving detection rates on intrusion detection systems. Expert Syst. Appl. **42**(1), 193–202 (2015)
27. Elkano, M., Galar, M., Sanz, J.A., Fernandez, A., Tartas, E.B., Herrera, F., Bustince, H.: Enhancing multiclass classification in farc-hd fuzzy classifier: on the synergy between n-dimensional overlap functions and decomposition strategies. IEEE Trans. Fuzzy Syst. **23**(5), 1562–1580 (2015)

28. Fazzolari, M., Alcala, R., Nojima, Y., Ishibuchi, H., Herrera, F.: A review of the application of multi-objective evolutionary systems: current status and further directions. IEEE Trans. Fuzzy Syst. **21**(1), 45–65 (2013)
29. Fernandez, A., Almansa, E., Herrera, F.: Chi-Spark-RS: an spark-built evolutionary fuzzy rule selection algorithm in imbalanced classification for big data problems (2017)
30. Fernandez, A., Carmona, C., del Jesus, M., Herrera, F.: A view on fuzzy systems for big data: progress and opportunities. Int. J. Comput. Intell. Syst. **9**(1), 69–80 (2016)
31. Fernández, A., Río, S., López, V., Bawakid, A., del Jesus, M.J., Benítez, J., Herrera, F.: Big data with cloud computing: an insight on the computing environment, MapReduce and programming framework. WIREs Data Mining Knowl. Discov. **4**(5), 380–409 (2014)
32. Fernandez, A., Altalhi, A., Alshomrani, S., Herrera, F.: Why linguistic fuzzy rule based classification systems perform well in big data applications? Int. J. Comput. Intell. Syst. **10**, 1211–1225 (2017)
33. Fernandez, A., Lopez, V., del Jesus, M.J., Herrera, F.: Revisiting evolutionary fuzzy systems: taxonomy, applications, new trends and challenges. Knowl. Based Syst. **80**, 109–121 (2015)
34. Fernandez, A., del Rio, S., Lopez, V., Bawakid, A., del Jesus, M.J., Benitez, J.M., Herrera, F.: Big data with cloud computing: an insight on the computing environment, MapReduce and programming frameworks. Wiley Interdisc. Rev.: Data Mining Knowl. Discov. **4**(5), 380–409 (2014)
35. Fernandez, A., Calderon, M., Barrenechea, E., Bustince, H., Herrera, F.: Solving multi-class problems with linguistic fuzzy rule based classification systems based on pairwise learning and preference relations. Fuzzy Sets Syst. **161**(23), 3064–3080 (2010)
36. Ferranti, A., Marcelloni, F., Segatori, A., Antonelli, M., Ducange, P.: A distributed approach to multi-objective evolutionary generation of fuzzy rule-based classifiers from big data. Inf. Sci. **415–416**, 319–340 (2017)
37. Florez, G., Bridges, S., Vaughn, R.: An improved algorithm for fuzzy data mining for intrusion detection. In: Proceedings of the 21st North American Fuzzy Information Processing Society Conference (NAFIPS'02). pp. 457–462. New Orleans, LA (2002)
38. Gacto, M.J., Alcala, R., Herrera, F.: Adaptation and application of multi-objective evolutionary algorithms for rule reduction and parameter tuning of fuzzy rule-based systems. Soft Comput. **13**(5), 419–436 (2009)
39. Gacto, M.J., Alcala, R., Herrera, F.: Interpretability of linguistic fuzzy rule-based systems: an overview of interpretability measures. Inf. Sci. **181**(20), 4340–4360 (2011)
40. Galar, M., Fernández, A., Barrenechea, E., Bustince, H., Herrera, F.: An overview of ensemble methods for binary classifiers in multi-class problems: experimental study on one-vs-one and one-vs-all schemes. Pattern Recogn. **44**(8), 1761–1776 (2011)
41. Goldberg, D.E.: Genetic Algorithms in Search, Optimization, and Machine Learning. Addison-Wesley Professional, Upper Saddle River, NJ, USA (1989)
42. Gomez, J., Dasgupta, D.: Evolving fuzzy classifiers for intrusion detection. In: Proceedings of IEEE Workshop on Information Assurance. pp. 68–75. United States Military Academy, West Point, New York (2001)
43. Gorzalczany, M., Rudzinski, F.: Interpretable and accurate medical data classification–A multi-objective genetic-fuzzy optimization approach. Expert Syst. Appl. **71**, 26–39 (2017)
44. Greene, D.P., Smith, S.F.: Competition-based induction of decision models from examples. Mach. Learn. **13**(2–3), 229–257 (1993)
45. Herrera, F.: Genetic fuzzy systems: taxonomy, current research trends and prospects. Evol. Intell. **1**(1), 27–46 (2008)
46. Herrera, F., Charte, F., Rivera, A.J., del Jesús, M.J.: Multilabel Classification-Problem Analysis. Springer, Metrics and Techniques (2016)
47. Herrera, F., Ventura, S., Bello, R., Cornelis, C., Zafra, A., Tarragó, D.S., Vluymans, S.: Multiple Instance Learning—Foundations and Algorithms. Springer (2016)
48. Holland, J.H.: Adaptation in Natural and Artificial Systems. University of Michigan Press, Ann Arbor, MI, USA (1975)

49. Homaifar, A., McCormick, E.: Simultaneous design of membership functions and rule sets for fuzzy controllers using genetic algorithms. IEEE Trans. Fuzzy Syst. **3**(2), 129–139 (1995)
50. Ishibuchi, H., Murata, T., Turksen, I.: Single-objective and two-objective genetic algorithms for selecting linguistic rules for pattern classification problems. Fuzzy Sets Syst. **8**(2), 135–150 (1997)
51. Ishibuchi, H., Nozaki, K., Yamamoto, N., Tanaka, H.: Selection of fuzzy IF-THEN rules for classification problems using genetic algorithms. IEEE Trans. Fuzzy Syst. **3**(3), 260–270 (1995)
52. Karnik, N.N., Mendel, J.M., Liang, Q.: Type-2 fuzzy logic systems. IEEE Trans. Fuzzy Syst. **7**(6), 643–658 (1999)
53. Khor, K.C., Ting, C.Y., Phon-Amnuaisuk, S.: A cascaded classifier approach for improving detection rates on rare attack categories in network intrusion detection. Appl. Intell. **36**(2), 320–329 (2012)
54. Kim, D., Choi, Y., Lee, S.Y.: An accurate cog defuzzifier design using lamarckian co-adaptation of learning and evolution. Fuzzy Sets Syst. **130**(2), 207–225 (2002)
55. Konar, A.: Computational intelligence: principles, techniques and applications. Springer, Berlin, Germany (2005)
56. Kuok, C.M., Fu, A.W.C., Wong, M.H.: Mining fuzzy association rules in databases. SIGMOD Rec. **27**(1), 41–46 (1998)
57. Lee, W., Stolfo, S.: A framework for constructing features and models for intrusion detection systems. ACM Trans. Inf. Syst. Secur. **3**(4), 227–261 (2000)
58. Liao, T.: A procedure for the generation of interval type-2 membership functions from data. Appl. Soft Comput. J. **52**, 925–936 (2017)
59. Marquez, F., Peregrín, A., Herrera, F.: Cooperative evolutionary learning of linguistic fuzzy rules and parametric aggregation connectors for mamdani fuzzy systems. IEEE Trans. Fuzzy Syst. **15**(6), 1162–1178 (2008)
60. Mohammadi Shanghooshabad, A., Saniee Abadeh, M.: Sifter: an approach for robust fuzzy rule set discovery. Soft Comput. **20**(8), 3303–3319 (2016)
61. Muhuri, P., Ashraf, Z., Lohani, Q.: Multi-objective reliability-redundancy allocation problem with interval type-2 fuzzy uncertainty. IEEE Trans, Fuzzy Syst (2017)
62. Naik, N., Diao, R., Shen, Q.: Dynamic fuzzy rule interpolation and its application to intrusion detection. IEEE Trans, Fuzzy Syst (2017)
63. Özyer, T., Alhajj, R., Barker, K.: Intrusion detection by integrating boosting genetic fuzzy classifier and data mining criteria for rule pre-screening. J. Netw. Comput. Appl. **30**(1), 99–113 (2007)
64. Patcha, A., Park, J.M.: An overview of anomaly detection techniques: Existing solutions and latest technological trends. Comput. Netw. **51**(12), 3448–3470 (2007)
65. Pedrycz, W., Gomide, F.: Fuzzy Systems Engineering: Toward Human-Centric Computing, 1st edn. Wiley (2007)
66. Quinlan, J.R.: C4.5: Programs for Machine Learning. Morgan Kaufmann Publishers, San Mateo-California, USA (1993)
67. Sambuc, R.: Function Φ-flous, application a l'aide au diagnostic en Pathologie Thyroidienne. Ph.D. thesis, University of Marseille (1975)
68. Rey, M., Galende, M., Fuente, M., Sainz-Palmero, G.: Multi-objective based fuzzy rule based systems (FRBSS) for trade-off improvement in accuracy and interpretability: a rule relevance point of view. Knowl. -Based Syst. **127**, 67–84 (2017)
69. Sanz, J.A., Fernandez, A., Bustince, H., Herrera, F.: Improving the performance of fuzzy rule-based classification systems with interval-valued fuzzy sets and genetic amplitude tuning. Inf. Sci. **180**(19), 3674–3685 (2010)
70. Sanz, J.A., Fernandez, A., Bustince, H., Herrera, F.: IVTURS: a linguistic fuzzy rule-based classification system based on a new interval-valued fuzzy reasoning method with tuning and rule selection. IEEE Trans. Fuzzy Syst. **21**(3), 399–411 (2013)
71. Sanz, J., Fernandez, A., Bustince, H., Herrera, F.: A genetic tuning to improve the performance of fuzzy rule-based classification systems with interval-valued fuzzy sets: degree of ignorance and lateral position. Int. J. Approx. Reasoning **52**(6), 751–766 (2011)

72. Smith, S.: A learning system based on genetic algorithms. Ph.D. thesis, University of Pittsburgh, Pittsburgh, PA (1980)
73. Smith, S.: Flexible learning of problem solving heuristics through adaptive search. In: 8th International Joint Conference on Artificial Intelligence, pp. 422–425 (1983)
74. Tajbakhsh, A., Rahmati, M., Mirzaei, A.: Intrusion detection using fuzzy association rules. Appl. Soft Comput. **9**(2), 462–469 (2009)
75. Thrift, P.: Fuzzy logic synthesis with genetic algorithms. In: Proceedings of the 4th International Conference on Genetic Algorithms (ICGA'91), pp. 509–513 (1991)
76. Tsakiridis, N., Theocharis, J., Zalidis, G.: DECO3RUM: a differential evolution learning approach for generating compact mamdani fuzzy rule-based models. Expert Syst. Appl. **83**, 257–272 (2017)
77. Tsang, C.H., Kwong, S., Wang, H.: Genetic-fuzzy rule mining approach and evaluation of feature selection techniques for anomaly intrusion detection. Pattern Recogn. **40**(9), 2373–2391 (2007)
78. Vasilomanolakis, E., Karuppayah, S., Muhlhauser, M., Fischer, M.: Taxonomy and survey of collaborative intrusion detection. ACM Comput. Surv. 47(4), 55:1–55:33 (2015)
79. Venturini, G.: SIA: a supervised inductive algorithm with genetic search for learning attributes based concepts. In: Brazdil, P. (ed.) Machine Learning ECML–93. LNAI, vol. 667, pp. 280–296. Springer (1993)
80. Victorie, T.A., Sakthivel, M.: A local search guided differential evolution algorithm based fuzzy classifier for intrusion detection in computer networks. Int. J. Soft Comput. **6**(5–6), 158–167 (2012)
81. Wang, H., Kwong, S., Jin, Y., Wei, W., Man, K.F.: Agent-based evolutionary approach for interpretable rule-based knowledge extraction. IEEE Trans. Syst. Man Cybernet. Part C: Appl. Rev. **35**(2), 143–155 (2005)
82. Wu, S.X., Banzhaf, W.: The use of computational intelligence in intrusion detection systems: a review. Appl. Soft Comput. **10**(1), 1–35 (2010)
83. Yager, R.R., Filev, D.P.: Essentials of fuzzy modeling and control. Wiley (1994)
84. Zadeh, L.A.: Fuzzy sets. Inf. Control **8**, 338–353 (1965)
85. Zarpelao, B., Miani, R., Kawakani, C., de Alvarenga, S.: A survey of intrusion detection in internet of things. J. Netw. Comput. Appl. **84**, 25–37 (2017)
86. Zhu, D., Premkumar, G., Zhang, X., Chu, C.H.: Data mining for network intrusion detection: a comparison of alternative methods. Decis. Sci. **32**(4), 635–660 (2001)

Printed in the United States
By Bookmasters